Synthesis and Tribological Applications of Hybrid Materials

# Synthesis and Tribological Applications of Hybrid Materials

*Edited by Mohammad Jawaid, Rajini Nagarajan, Jacob Sukumaran, and Patrick De Baets*

WILEY-VCH

**Editors**

***Dr. Mohammad Jawaid***
Universiti Putra Malaysia
Institute of Tropical Forestry and Forest Products (INTROP)
Serdang
43400 Selangor
Malaysia

***Dr. Rajini Nagarajan***
Kalasaingam University
Dept. of Mechanical Engineering
Anand Nagar
626126 Krishnankoil, Tamil Nadu
India

***Dr. Jacob Sukumaran***
Ghent University
Laboratory Soete
Technologiepark Zwijnaarde 903
9052 Zwijnaarde
Belgium

***Prof. Patrick De Baets***
Ghent University
Dept. Electr. Energy, Systems, Automat.
Sint-Pietersnieuwstraat 25
9000 Gent
Belgium

**Cover:** © nikkytok/Shutterstock



**Library of Congress Card No.:** applied for

**British Library Cataloguing-in-Publication Data**
A catalogue record for this book is available from the British Library.

**Bibliographic information published by the Deutsche Nationalbibliothek**
The Deutsche Nationalbibliothek lists this publication in the Deutsche Nationalbibliografie; detailed bibliographic data are available on the Internet at <http://dnb.d-nb.de>.



**Print ISBN:** 978-3-527-34301-0
**ePDF ISBN:** 978-3-527-80861-8
**ePub ISBN:** 978-3-527-80859-5
**oBook ISBN:** 978-3-527-80858-8

**Typesetting** SPi Global, Chennai, India
**Printing and Binding** C.O.S. Printers Pte Ltd, Singapore

Printed on acid-free paper

10 9 8 7 6 5 4 3 2 1

# Contents

# Preface

Nowadays, from day-to-day activities of human life to high-end sophisticated applications, tribology has a role to play in any one of the aspects. Hence, the need to understand the process, materials, techniques, and mechanism related to tribological concepts is as an important factor to be considered by the research community. Moreover, tribology is a system-dependent phenomenon that covers a wide range of science and technology fields including materials science, mechanical engineering, lubricant engineering, biomaterials, chemistry, physics, biology, and many more, encompassing micro as well as nanotechnology.

Major organizations and academic institutions in developing countries have also realized the importance of industrial tribology and its relevance to modern design and maintenance as a result of immense savings on declining material and energy losses and decrease in cost of mass production. Accordingly, the design of eco-friendly material for the tribological components is found to be the primary factor in the energy saving sector as well as for a green environment. One such material system is the bio-degradable natural fiber reinforced self-lubricating polymer composite, which can be a suitable alternative for external lubricant tribo systems. Nowadays, the research advancement of degradable natural fiber reinforced polymer composites in the field of automotive, aerospace, and construction industries has been appreciable because it provides much economic benefit due to its inherent quality of higher specific strength to weight ratio. Although the structural properties are important for tribological applications, priority will have to be given for enhancing the wear and friction behavior. Natural fiber based polymer composites have attracted researchers attention for tribological applications because of their excellent structural performance and have excelled over conventional materials in industrial systems such as bearings, automobile parts, gears, and chute liners while the advantage of reinforcing two fiber materials in one matrix or one fiber in polymer blends through hybridization effect. The development of hybrid composites helps attain a combination of the exceptional mechanical properties of different fibers/fillers. Scientists have also investigated the tribological characteristics of natural fiber composites hybridized with secondary fillers with varying operational parameters such as the type of fillers, grid size, and wt % matrix. Generally, it has been found that nano- or micro-sized inorganic particles can significantly influence the tribological properties of polymeric materials.

This book covers versatile topics such as tribological assessment on accelerated aging bones in polymeric condition, nano fracture and wear testing on natural bones, tribological behavior of glass fiber with fillers reinforced hybrid polymer composites and jute/glass hybrid composites, wear properties of glass fiber hybrid, and acid- and silane-modified CNT filled hybrid glass/kenaf epoxy composites, hybrid natural fiber composites as a friction material, comparative wear model on hybrid natural fiber composites, tribological behavior of particulate and carbon fiber reinforced epoxy hybrid composites, and dry sliding wear behavior of copper based hybrid metal matrix composites.

We are highly thankful to all authors who contributed the chapters and provided their valuable ideas and knowledge on tribology in this edited book. We have attempted to gather all the information from recognized researchers from Malaysia, India, Brazil, and Hungary in the areas of tribological properties and have finally completed this venture in a fruitful manner. We greatly appreciate the contributors' commitment and their support to compiling our ideas in reality.

We are highly thankful to the Wiley team for their generous cooperation at every stage of the book production.

*Mohammad Jawaid, Malaysia*
*Rajini Nagarajan, India*
*Jacob Sukumaran, Belgium*
*Patrick de Baets, Belgium*

# 1

# Tribological Assessment on Accelerated Aging Bones in Polymeric Condition

*Ramdziah M. Nasir, Law C. Gan, and Abdul Y. Saad*

*University of Science, Malaysia, School of Mechanical Engineering, Engineering Campus, Nibong Tebal 13400, Penang, Malaysia*

## 1.1 Introduction

The global trends of doctor-diagnosed arthritis and osteoporosis show a dramatic increase. According to a statistical survey by the Centers for Disease Control and Prevention (CDC), the percentage of adults suffering this disease has gradually increased for both women and men, especially the aged, as shown in Figures 1.1 and 1.2 [1]. Lack of exercise in daily life, obesity, and family-inherited genetic problems are the main causes of the disease.

Recently, there has been much effort by researchers toward improving bone defects – for example, the transplant of scaffold for regeneration of new bones. However, most of the research is focused on the strengthening of defective bones by using the replacement method. There is a possibility to research on transplant of stronger bones from one person to another in the near future.

The first stage is to observe the changes in properties of stronger animal bones at different conditions that include temperature and relative humidity (RH). Research on human bones can help determine the factors that can strengthen the physical properties and enhance the elemental properties of bones. This project will study animal bones such as bovine and goat bones for strength and simulate the similarity between human bones and animal bones in terms of accelerating aging factors and conditions. The work will focus on selecting the most suitable animal model for the study to simulate human bones and to assess how daily activity and load bearing affect the properties of animal bones. Parameters such as temperature, hardness, and tensile forces that affect the properties of animal bones are assessed by comparing the structural changes of two samples of bone before and after exposure to different conditions.

*Synthesis and Tribological Applications of Hybrid Materials*, First Edition.
Edited by Mohammad Jawaid, Rajini Nagarajan, Jacob Sukumaran, and Patrick De Baets.

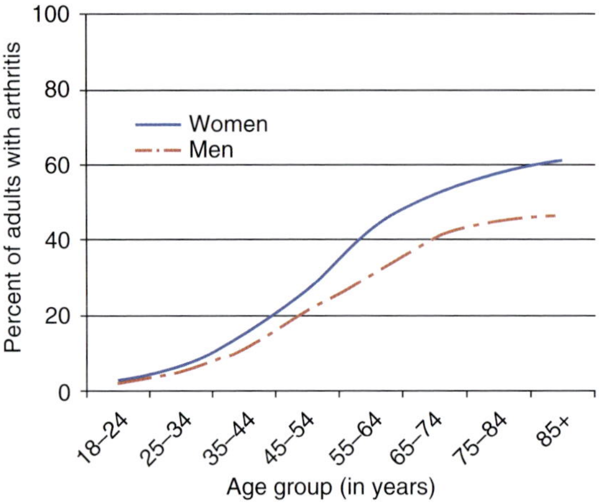


**Figure 1.1** Sex-specific prevalence of doctor-diagnosed arthritis, National Health interview, from 2007–2009 [1].

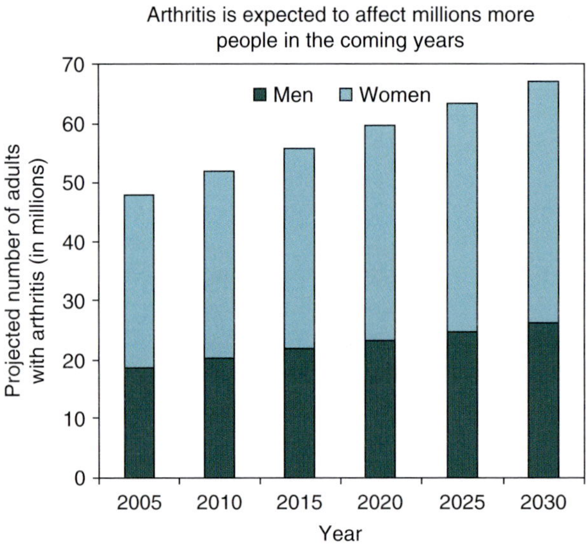


**Figure 1.2** Projected prevalence of doctor-diagnosed arthritis among adults 2005–2030 [1].

## 1.2 Bone

In order to determine how the various environmental factors can affect the properties of the selected animal bone model, the basic knowledge of the composition of bones must be revisited. Much work has been done to explore how the various types of bones react in various environments. Bone is a composite that consists of minerals (calcium phosphate or hydroxyapatite) and protein (collagen) in a hierarchical architecture [2]. Approximately 20% of bone consists of water;

35% of the dry bone material consists of collagen, proteins, and glycosaminoglycans [3] and the remaining consists of other hybrids of organic and inorganic substances.

Basically, bone not only provides robust mechanical support for the whole human body but also protects the various organs from external impacts. The mechanical properties of bone are good owing to its strong structural hierarchical architecture in the human body, built with strong and stiff minerals and weak and soft proteins [2]. In the past, Isaksson et al. [4] reported that majority of mammal bones are similar to one another except for the presence of plexiform bone in large mammals such as pigs, cows, and goats [5]. Plexiform bone is primarily found in large rapidly growing animals and offers increased mechanical support for long periods of time. It is reported by Pearce et al. [6] that there are a few animal models that are suitable for use in testing of orthopedic and dental implants prior to clinical use in humans. Frequently used bones include dog, sheep, goat, pig, and rabbit models for the evaluation of bone–implant interactions. There were some minor differences in bone composition between the various species and humans. The pig has a good likeness with human bone but difficulties arise due to its size and ease of handling. So, the dog and goat were chosen as better animal models for testing of bone implant materials. Moreover, Pearce et al. [6] had proved that goat is the chosen species for 8.2% of animal studies published in the Journal of Orthopedic Research between 1992 and 1996. In comparison with sheep, goats tend to have a more inquisitive and interactive nature, which can make long time confinement easier than with sheep. Besides, in the Southeast Asian region, which has high temperature and humidity, goats are more tolerant toward ambient conditions than other species. According to biomechanical consideration of animal models used in tissue engineering, a small difference is apparent in the ashes density between goats and humans. This indicates that there is not much variation between anatomic sites of the same species. Besides, the goat has the same mineral composition as humans. Prothero et al. [7] had studied the attributes of hoofed mammals whereby the tarsal of the foot of the cow can be compared to the human foot. At this point, they had started to observe the structural and functional similarities between cow's leg and the human foot.

According to ASTM E8, strip materials commonly used for tensile testing are in dog bone shape, with wide ends and a narrow middle [8]. According to ASTM [9], conducting the experiment and performing the milling process by using bone samples of "dog bone" shape for the bone-shaped tensile specimen can lower the ultimate force at failure, which can reduce the stress concentration in the grips and concentrate failure in specific area for brittle mass. Besides, Subit et al. [10] had discovered that the gradual taper allows for a smooth transition of load distribution from the edge to the center. So, it can reduce the possibility of stress concentration from the sample edge and focus the failure in the gauge section of the sample. There were a few studies by the researchers on RH testing. Ajadi and Sanusi [11], Huynh et al. [12], and Karr and Outram [13] had studied the effect of RH on oven temperature of solar cabinet dryer. There were a few studies on tensile tests on bones [14–22].

In general, [22–25] have conducted nano-indentation on bovine specimen surface, with a surface contact force of 30 mN and a constant loading rate of 2000 μN $s^{-1}$. The hardness depression was held for a period of 5 s at the maximum load to eliminate creep behavior, before unloading at the rate of 2000 μN $s^{-1}$. Rho et al. [26] also have performed nano-indentation using nano-indenter II, which is a fully automated hardness testing system. The load and displacements of the indenter used were 0.3 μN and 0.16 nm respectively. Sharp Berkovich diamond indenters, a three-sided pyramid with the same area-to-depth ratio as the Vickers indenter, were also used. They reassessed the elastic modulus and hardness of secondary osteonal and interstitial bone through the cortex of human femora of various ages using the nano-indenter [27]. To minimize the effects of viscoelasticity and creep on proper measurement, a long constant load hold period was introduced before final loading. The second constant load hold period was introduced near the end of the test, at 10% of the peak load, in order to establish the rate of thermal expansion or contraction of the testing apparatus to correct the displacement data for thermal drift. The nano-indentation was conducted with a maximum load of 8 mN, with a loading rate of 400 μN $s^{-1}$, which can produce a hardness impression with a depth of 700 nm. Three target areas from near the center of the specimen along the radial direction were selected in each endosteal, middle, and periosteal site of the specimens. Van had also determined the hardness of the cortical bone in goats at different preservation durations by conducting the Vicker test. In the experiment, a Vicker automatic hardness measurement device (Buehler 1600-6400; Buehler, Lake Bluff, IL) was used to perform six measurements on each specimen [28]. The indentation was made using a diamond indenter and an indent weight of 50 g. Then, the Ominimet MHT software was used to analyze the indentation.

## 1.3 Methodology

The experimental methodology was divided into four phases. They include planning the parameters selection for the experiment, design of the experiment, observation, and analysis phases. Design of experiment (DOE) plays an important role in optimization of the temperature parameter in accelerated temperature testing. In order to determine the desired information, both the number of factors and levels of parameters must be predetermined.

### 1.3.1 Phase I: Planning

Before carrying out any experiment, the parameters for each experiment were studied and discussed. The information gained from journals on the experiments done by previous researchers, suggestions from the supervisor and laboratory technicians, and references from standards were all recorded for DOE settings.

After analyzing the information, decision making on the parameter settings for the experiments was carried out. In accelerated temperature testing, the parameters used were temperature and RH. The RH values in the closed oven depended on the temperature and were calculated based on the difference in the temperatures of the wet and dry thermocouples. The maximum load (*N*) and hardness values of each specimen were compared between bovine and goat bones in mechanical testing and using nano-indentation respectively.

### 1.3.2 Phase II: Design of Experiment

In this study, the accelerated temperature testing has two dependent variables, temperature and RH. In the DOE, the temperature was set to three levels (low, medium, and high levels). Low level temperature was set at 22 °C, medium temperature at 55 °C, and high level temperature at 85 °C. Some bovine and goat bone specimens were tested similarly with each combination of temperature and RH. The accelerated temperature test was carried out for a duration of 10 h.

While RH was calculated based on the difference between the wet and dry thermocouple readings (T1 and T2 channel reading on thermometer) and by referring to the humidity chart (Appendix 1.A). The average RH values ($a$, $b$, and $c$) were obtained for each set of experiment.

The independent variables were determined using the tensile test and hardness test. By referring to the ASTM International Standard [9], the specific standard was set for strain rate for the tensile test, and a constant load and constant load rate were set for the nano-indentation test. The results of tensile test ($j$, $l$, $m$) and nano-indentation test ($x$, $y$, $z$) were analyzed and compared for both bovine and goat bones. The parameters design for each testing is illustrated in the Table 1.1.

### 1.3.3 Phase III: Conduct of Experiment

Goat and bovine bones were selected as the specimens for testing as they have similar properties to human bones. Firstly, the bones from cow and goat remains were collected. Then, these bones were dipped in water, boiled thoroughly, and flipped occasionally using a stirrer in the hot water pot for 4 h before the flesh was removed using a knife. The processes are shown in Figure 1.3.

On completion of cleaning, the tray was placed under the hot sun for 10 h daily and dried for 30 days. In between, the bones were frozen at −4 °C to eliminate bacterial attack and fungus. After drying, these bones were wrapped with a transparent film and kept in a small desiccator in a freezer.

**Table 1.1** Design of experiment.

| | Levels | | |
|---|---|---|---|
| **Condition** | **Low** | **Medium** | **High** |
| Temperature (°C) | 20 | 50 | 80 |
| Relative humidity (RH) | $a$ | $b$ | $c$ |
| Tensile test | $j$ | $l$ | $m$ |
| Hardness test (nano-indentation test) | $x$ | $y$ | $z$ |

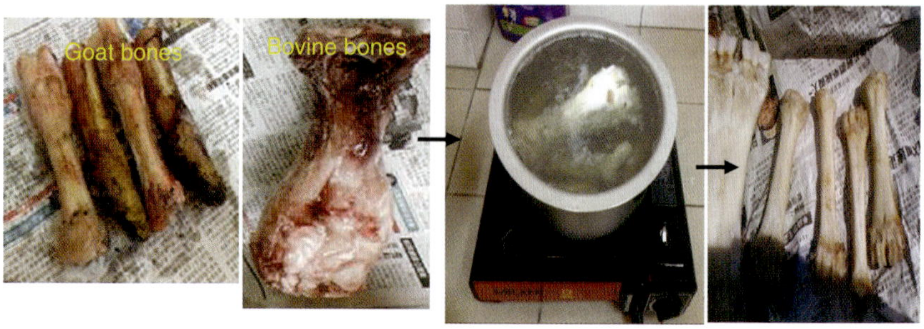


**Figure 1.3** Bone processing from raw bone to boiling and de-fleshing.

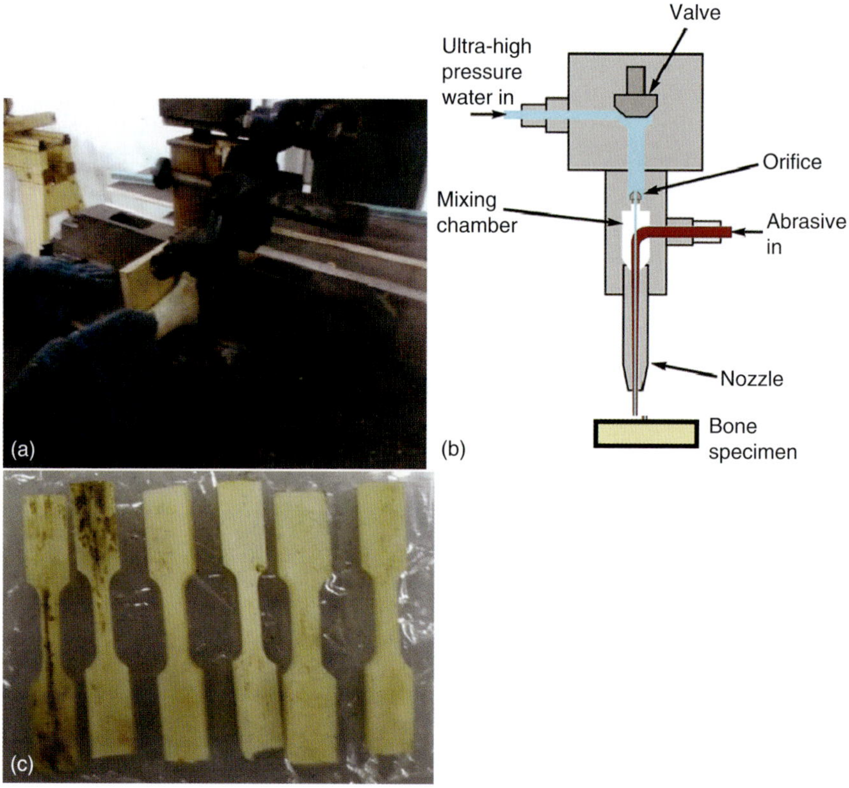


**Figure 1.4** Cutting process: (a) manual band saw, (b) abrasive water jet, and (c) dog bone shaped specimen.

Prior to fabrication, the bones were cut along the longitudinal axis by using an abrasive water jet band saw as shown in Figure 1.4, cleaned, and polished using sandpaper and profiled. Then, the bone was ground into rectangular shape using a grinder machine for easier machining purposes later.

Afterwards, the bones were cut into dog bone shape through the abrasive water jet cutting process as shown in Figure 1.4b. In the process of abrasive water jet cutting, the high pressure pump produced a high pressure and directed the pressurized water from the pump to the cutting head through the high pressure supplies line. When the pressurized water came out from the nozzle, an extremely high velocity water jet was created. The solid abrasive particles were then mixed with the water jet. The resulting high-speed abrasive water jet was focused on the bone material through the abrasive nozzle. Then, the kinetic energy of the abrasive particles was used to remove unwanted areas on the bones. The finished dog bone shaped specimen was prepared as shown in Figure 1.4c.

The dog-shaped bones were then cleaned in ultrasonic machine for 3 min before being hot dried in the oven for 5 min. These initial bone structures were observed under scanning electron microscope (SEM) and the composition information analyzed under energy-dispersive X-ray analysis (EDAX). The procedures of SEM/EDAX inspection will be discussed in detail in phase III.

Thirty grams of epoxy resin and 10 g of curing agent HB-2215 were mixed and stirred evenly for 10 min. The goat and cow bones were dipped into the solution separately for 5 min. These bones were then removed and put on a non-sticky plastic sheet. After completing all deep coating processes on bones, they were cured at room temperature to avoid bubbles. Then, the non-sticky plastic sheet together with bones was dried in a drying cabinet for 8 h at a temperature of 40 °C. After 8 h, the plastic sheet was removed. All the bones were put into transparent packaging separately. These bones were observed again under SEM/EDAX.

In the accelerated temperature testing, a laboratory oven was used. A wet–dry bulb hygrometer concept was used in the experiment. Two thermocouples and a digital thermometer were prepared. One of the thermocouples was tied with small pieces of wet cotton. Both thermocouples were connected to the digital thermometer channels (T1 and T2) separately. The laboratory oven was turned on and set to the desired temperature (55 and 85 °C). After reaching the stable temperature, the bovine and goat bones were put together inside the oven. The readings of T1 and T2 channels were taken every 15 min. It is important to take the exact reading for both T1 and T2 channels when they have reached the stable state. The experiment was carried out for 7 h and a total of 28 readings were taken. The RH in the laboratory oven was calculated based on the humidity chart. The setup of temperature measurement is illustrated in Figure 1.5.

The tensile test was performed using Instron Table Mounted Universal Tensile Machine (UTM) with a 30 kN load cell. Before carrying out tensile test, the gauge length, width, and thickness of each bone was measured five times and the average value was obtained. Each specimen has a different thickness and width but same gauge length as shown in Table 1.2. The first bone sample was mounted on the grip and safely secured by closing the grips. The test lasted for about 1 min. For strain evaluation, the cross-head speed was set at 0.025 mm $s^{-1}$. All the tests for different bone specimens used the same cross-head speed. The gauge length was set at 10.5 mm apart.

After the tensile test, the bone specimens were tested for their hardness properties by using Micro Materials Nano Test nano-indenter. The machine offers a complete range of nanomechanical and nanotribological tests in a flexible and user-friendly instrument. The bone samples to be measured must be ideally flat

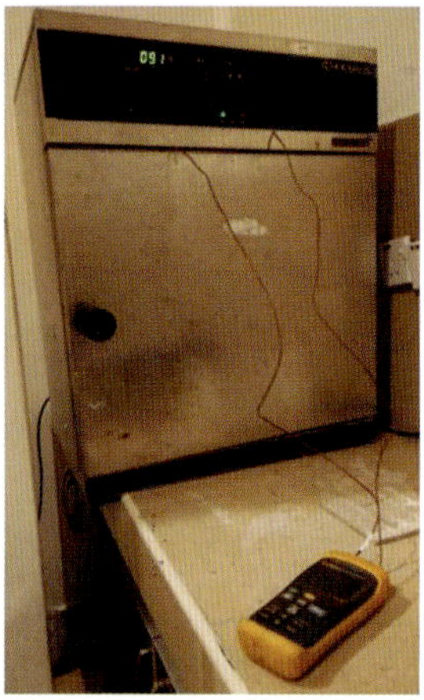

Figure 1.5 Temperature measurement using wet–dry bulb hygrometer concept.

Table 1.2 The dimension of both bovine and goat specimens.

| Specimen | Average thickness (mm) | Average width (mm) | Gauge length (mm) |
|---|---|---|---|
| B3 | 4.144 | 2.944 | 10.500 |
| B4 | 3.308 | 3.114 | 10.500 |
| B2 | 4.110 | 2.992 | 10.500 |
| G3 | 1.986 | 2.892 | 10.500 |
| G4 | 2.648 | 3.006 | 10.500 |
| G2 | 2.126 | 2.762 | 10.500 |

B: Bovine bone, G: goat bone.

with fine surface finish. Then, the specimen was mounted on the stub using a small quantity of cyanoacrylate adhesive.

### 1.3.4 Phase IV: Observation and Analysis

In order to inspect the microstructure inside the bone, the SEM Hitachi S-3400N was used. SEM has a better resolution of 10 nm at 3 kV and it is computer eucentric, which has a five-axes motorized stage with −20°/+90° tilt. Before the inspection is done using SEM machine, the bone specimen was coated with a conductive layer of coating in the sputtering unit. Besides, SEM is an analytical specimen chamber with an optimum geometry that is used for simultaneous accommodation of analytical techniques such as energy-dispersive X-ray spectroscopy (EDS), wavelength-dispersive spectroscopy (WDS) system, and electron backscatter diffraction (EBSD) analysis. By integrating these three analytical techniques on a single platform, it can provide a complete materials characterization solution, which is called EDAX.

## 1.4 Results and Discussion

### 1.4.1 Accelerated Weathering Study

Before carrying out the accelerated weathering study, the bovine and goat bones were prepared. The appearance of both bones with fleshes before cooking and without fleshes after cooking and removing of fleshes is shown in Table 1.3. Boiling process is done to decompose the soft muscle tissue remaining on the bone surface.

After cooking and removing the fleshes, the length and width of the bone changed and the bone became smaller. The bone color changed from totally white during cooking to slightly yellowish after being dried up.

During the accelerated aging, there was a significant change in the mass of bone after 10 h. There was not much difference in the appearance of the bone such as color changes or surface fracture before and after the experiment. The reason is that the duration of aging is short and the temperature is low and it does not bear much effect on the bone. This is quite a different result compared to previous work. Previously, Ubelaker [29] proved that the color of bone

**Table 1.3** Bone appearance before and after removing fleshes.

| Condition | Bovine | Goat |
|---|---|---|
| Before cooking | 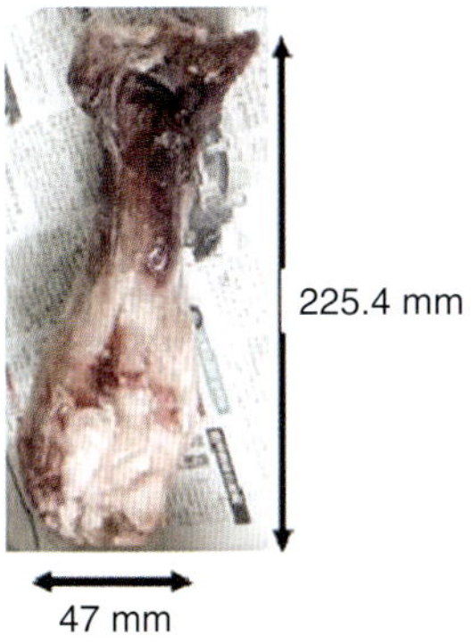  | 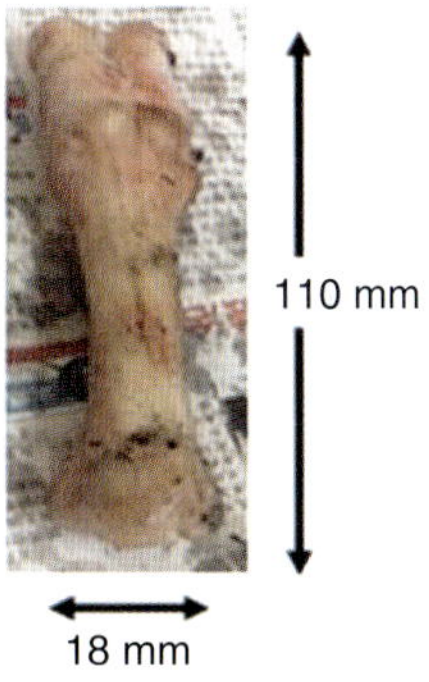  |
| After cooking and defleshing | 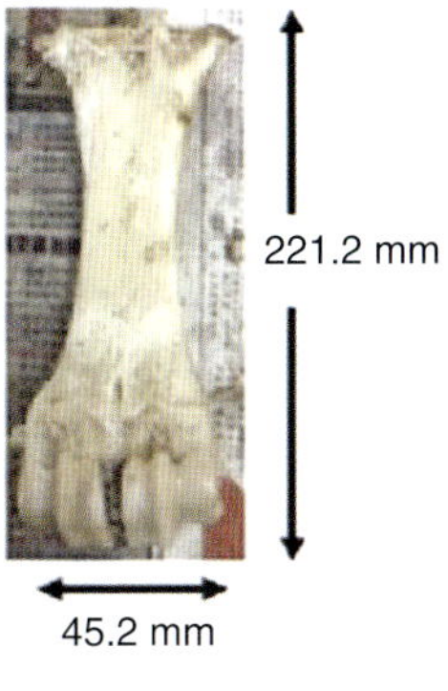  | 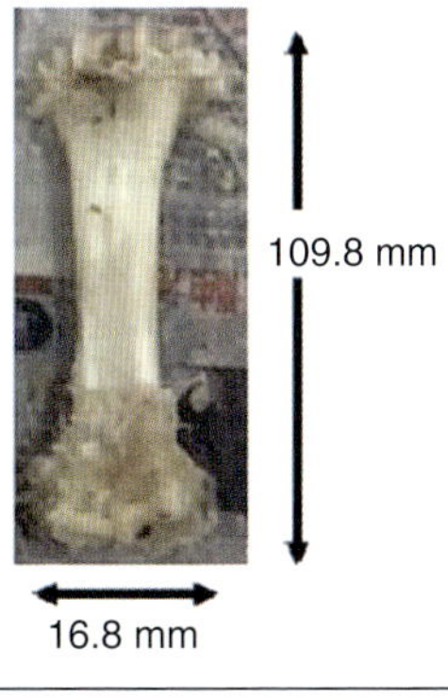  |

fragments is affected by high temperature as a function of duration and temperature, and oxygen availability. Most experiments done previously have shown that fresh bone follows the trend from brown through black, then gray–blue–white to white with increased temperature [30].

The first set of bovine and goat samples were placed at room temperature around 22 °C. The second and third sets of samples were placed in the laboratory oven at temperatures of 55 and 85 °C respectively. The RH at different temperatures is plotted in Figure 1.6.

There is one assumption made, that is, the RH of bovine and goat bones at room temperature is assumed to be around 60% [12, 31]. The first desired temperature was chosen to be 22 °C. It is assumed the bones undergo aging due to weathering at room temperature. Another sample was tested at the desired temperature, that is, 85 °C, under dry condition. The reason is that micro-organisms (thermophilic microbes) play a role in decomposition of bodies as they can live up to 80 °C. The last temperature chosen was 55 °C, which is the average between 22 and 85 °C.

From the graph in Figure 1.6, it can be seen clearly that RH has different values at different temperatures. Increasing the temperature will lower the RH while lowering the temperature will increase the RH. RH varies significantly when the temperature changes, even when the actual amount of water vapor is constant. In this case, lowering the temperature can result in a rise in RH even though there is no water source or water vapor added into the laboratory oven while heating the bones. The amount of water vapor present at saturation is less at lower temperatures. So, there is a higher percentage of saturation air for the existing amount of water vapor in the oven.

The humidity of weathering in the atmosphere fluctuates from day to day in tropical countries. The temperature and RH are recorded for day 1 (25 °C, 84% RH – dry day) and day 2 (26 °C, 84% RH – cloudy day). All the climatic variables such as temperature and humidity can affect each other. The change in temperature can cause a change in water evaporation and air saturation, and can directly lead to change in humidity.

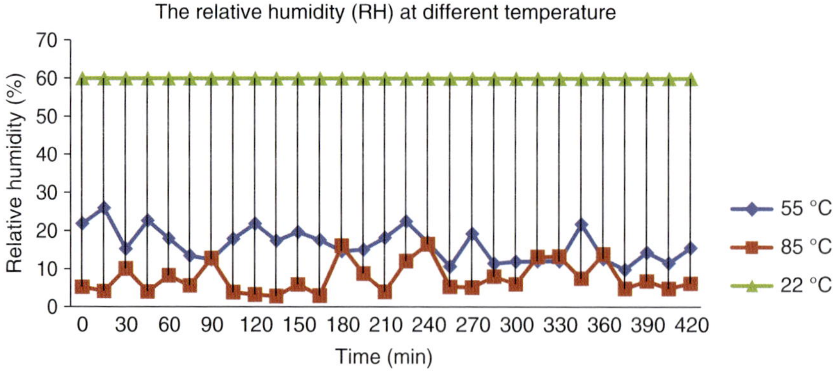


**Figure 1.6** Relative humidity versus time at different temperatures.

Besides, the air temperature differences between different locations can cause air pressure differences, which can produce different air movements in the oven. The initial reading of wet and dry thermocouples is different for every temperature reading taken. When both thermocouples are inserted into the oven and the temperature reading at different points measured, the final temperature reading taken is different. The principle is likely similar to the statement by Heritage [32].

However, there was a significant change in the mass of the bone after 10 h of accelerated aging. The mass of the bone specimen was measured before and after the accelerated aging process. There was not much difference in the appearance of bone such as color or cracks but the mass changed dramatically. A comparison of the mass loss is made in Table 1.4.

Based on Table 1.4, it can be seen that the maximum mass loss occurred in B3 specimen (2.256%) and minimum mass loss occurred in B1 specimen (0.099%). There is an increasing trend of mass loss when the temperature is increased.

Accelerated aging leads to changes in bone cells and tissues. These changes result in changes in bone mass due to maceration of bone. Maceration is a process where a clean skeleton is obtained from the animal carcass inside a closed and ventilated container. During the process, decomposition occurs, whereby the protein cells in the bone are broken down and consumed by bacteria. The mass of the bone reduces after exposure to accelerated aging conditions. The density of the bone is reduced with mass loss. The bone will become less dense. Thus, the bone becomes weaker and has high risk of fracture or lifetime osteoporosis.

With respect to determining the change in characteristics due to accelerated aging of the bones, it is important to determine the mechanical properties such as tensile strength. There are two different observations obtained from the experiment. The first section will discuss the effects of increase of temperature on the tensile strength of both bovine and goat bones. The Section 1.4.2 discusses the effects of an increase in temperature based on the physical activity of the cow and the goat.

**Table 1.4** Comparison of mass loss for bovine and goat specimen.

| | Mass loss, $\Delta m$ (g) | | | | | |
|---|---|---|---|---|---|---|
| | Bovine bone | | | Goat bone | | |
| | B1 (22 °C) | B2 (55 °C) | B3 (85 °C) | G1 (22 °C) | G2 (55 °C) | G3 (85 °C) |
| Before accelerated aging (g) | 2.018 | 1.574 | 2.128 | 0.866 | 1.262 | 0.875 |
| After accelerated aging (g) | 2.016 | 1,549 | 2.08 | 0.864 | 1.239 | 0.858 |
| Mass loss percentage (wt%) | 0.099 | 1.588 | 2.256 | 0.231 | 1.823 | 1.943 |

B: Bovine, G: goat.

### 1.4.2 Effects of Increase in Temperature and Decrease in Relative Humidity (RH) on Maximum Load Withstood by Bone

As the bovine and goat bone specimens were aged at increased temperatures, there was a significant change in tensile strength. When the bone specimens were aged at room temperature around 22 °C, the maximum load withstood by the bovine and goat bones (B3 and G3) was 667.41 and 713.06 N respectively as shown in Figures 1.7 and 1.8.

It shows that the goat bone can withstand a higher load than the bovine bone specimens. This indicates that the goat bone has higher strength than bovine bone under normal weathering condition at 22 °C.

When another set of bovine and goat bone specimens were heated at a temperature of 55 °C with a tolerance of ±5 °C, the maximum load withstood by the bovine and goat bone specimens (B4 and G4) increased to 958.18 and 1031.88 N respectively as shown in Figures 1.9 and 1.10.

The maximum tensile strength of bovine bone increased from 667.41 to 958.18 N. Similarly, the tensile strength of goat bone also increased from 713.06

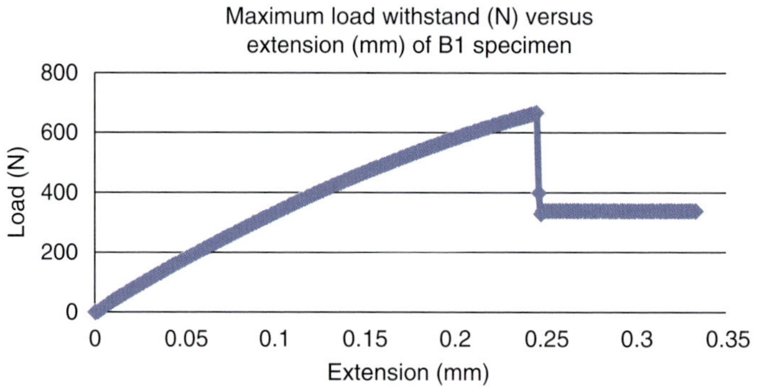


**Figure 1.7** Maximum load withstood (N) versus extension (mm) for bovine bone at a temperature of 22 °C.

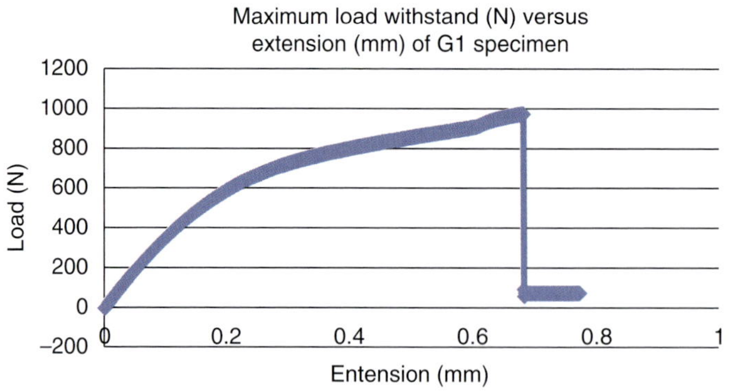


**Figure 1.8** Maximum load withstood (N) versus extension (mm) for goat bone at a temperature of 22 °C.

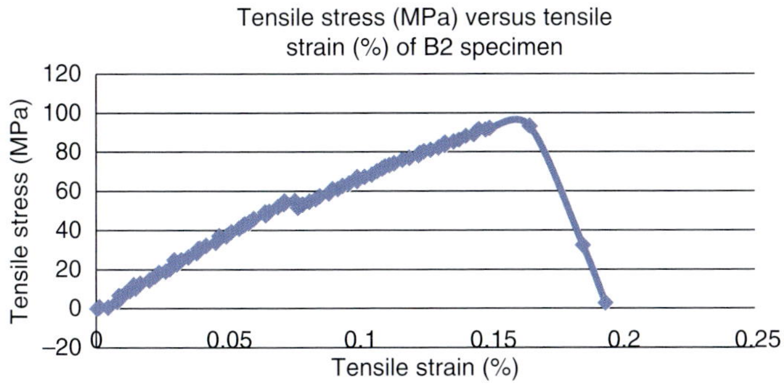


**Figure 1.9** Maximum load withstood (N) versus extension (mm) for bovine bone at a temperature of 55 °C.

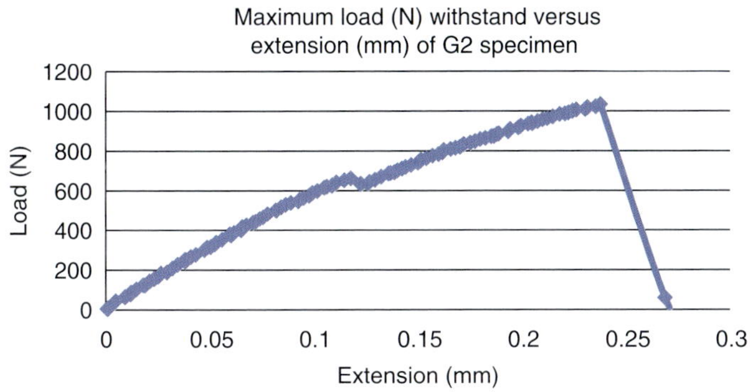


**Figure 1.10** Maximum load withstood (N) versus extension (mm) for goat bone at a temperature of 55 °C.

to 1031.88 N. When the bovine and goat specimens were aged at a temperature of 85 °C with a tolerance of ±5 °C, the maximum load withstood by the bovine and goat bone specimens (B2 and G2) was 501.64 and 713.06 N respectively as shown in Figures 1.11 and 1.12.

By comparison, the maximum tensile strength of bovine bone aged at 85 °C was lower than that of bovine bone aged at 55 °C. Again, the goat bone exhibited superior capability to withstand a maximum load of 713.06 N at 85 °C.

All the maximum loads that can be withstood by both bovine and goat bones are summarized in Table 1.5 and illustrated as a bar chart in Figure 1.13.

Based on Table 1.5, both bovine and goat bones have the maximum tensile strength at a temperature of 55 °C. This is followed by the bone specimen aging at 22 °C and the lowest loading at 85 °C. From this observation, it can said that the optimum aging of bone for improving the tensile strength is at 50 °C with an average RH of 15.08%.

Based on Figure 1.13, it can be seen that bovine and goat bones can withstand an average load at room temperature (low temperature, high RH). When aging at medium temperature (55 °C) and relatively medium RH, the bone can withstand

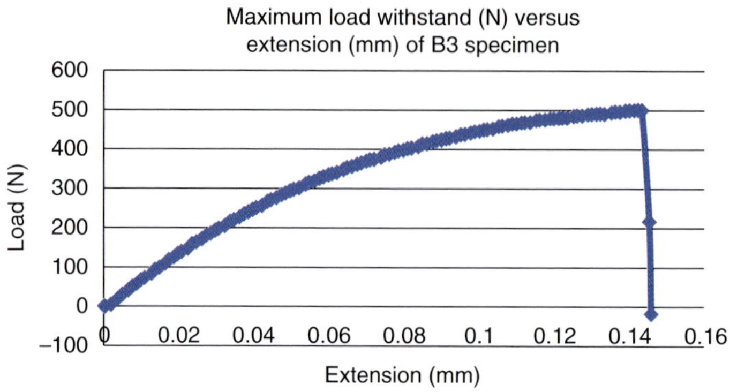


**Figure 1.11** Maximum load withstood (N) versus extension (mm) for bovine bone at a temperature of 85 °C.

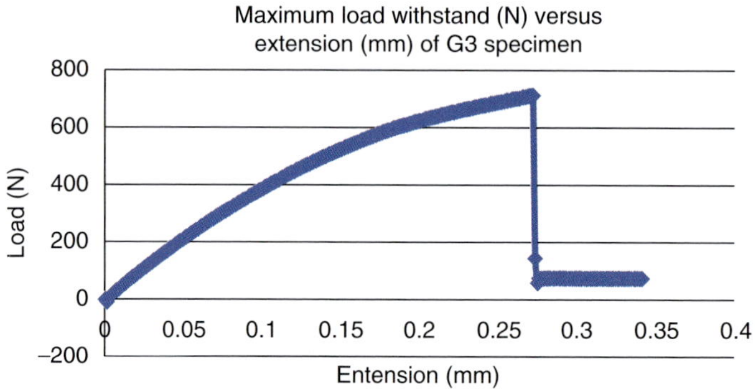


**Figure 1.12** Maximum load withstood (N) versus extension (mm) for goat bone at a temperature of 85 °C.

**Table 1.5** Comparison of maximum load withstood by bovine and goat bones at different temperatures.

| | Maximum load withstood (N) | |
|---|---|---|
| **Temperature/average relative humidity (RH)** | **Bovine** | **Goat** |
| 22 °C (60 ± 5% average RH) | 667.41 | 976.35 |
| 50 °C (15.08% average RH) | 958.18 | 1031.88 |
| 80 °C (7.39% average RH) | 501.64 | 713.06 |

a higher load before fracture. But, the bone aged at high temperature (85 °C) and low average RH is incapable of withstanding a higher tensile load. From the observation, it is clear that the bone has the strongest mechanical strength at moderate humidity and temperature. At relatively low temperatures, the average RH is high. So, it can create a suitable moist environment, and the wet bone is

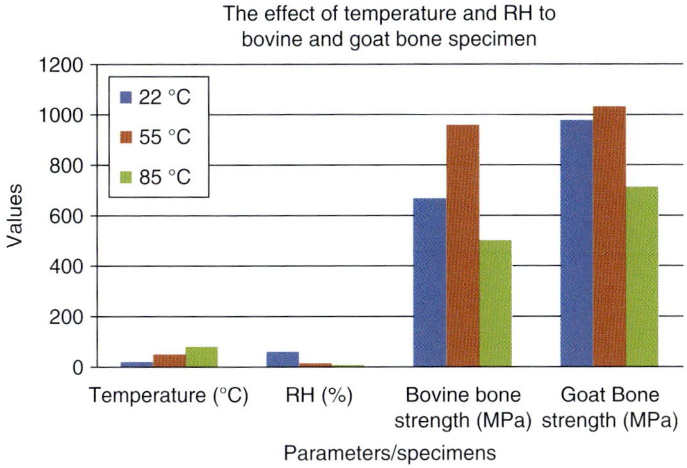


**Figure 1.13** The relationship of temperature and relative humidity (RH) to bovine and goat specimen strength.

highly susceptible to mold growth. Mold is a simple microscopic organism that destroys organic materials. The mold will digest the organic material in bone (collagen) and destroy it, and eventually spread to destroy the adjacent collagen network.

When a bone is aged at a very high temperature, bone deterioration occurs. The bone responds to the changes of RH by expanding and contracting to different degrees in longitudinal, radial, and tangential directions. During the aging process, the outer surface of the bone will dry and evaporate while the inner surface is still wet and expands internally. This may cause a big area to shrink, crack, and warp when the bone structure deteriorates gradually. Hence, it can affect the mechanical properties of the bone.

### 1.4.3 Comparison of Bovine and Goat Bone Strength at Normal and Increased Temperatures

From Figure 1.14, it can be seen that the goat bone has greater strength compared to the bovine bone at room temperature and at increased environmental temperature (55 °C). It proves that the bone structure of goat bone specimen is better than that of bovine bone specimen. It may be due to the effect that lifelong exercise has on the bone properties. The goat is predicted to perform more physical activity than the cow. More exercise can improve the properties of bone and its collagen network [4].

Even though goat bone size is smaller than that of bovine bone, it has greater mechanical properties. It may be due to porosity and stiffness. Porosity in structural materials exerts a strong influence on mechanical properties. The strength and stiffness vary inversely with increased porosity [18]. This indicates that the microstructure in goat bone has a more compacted arrangement compared to bovine bone.

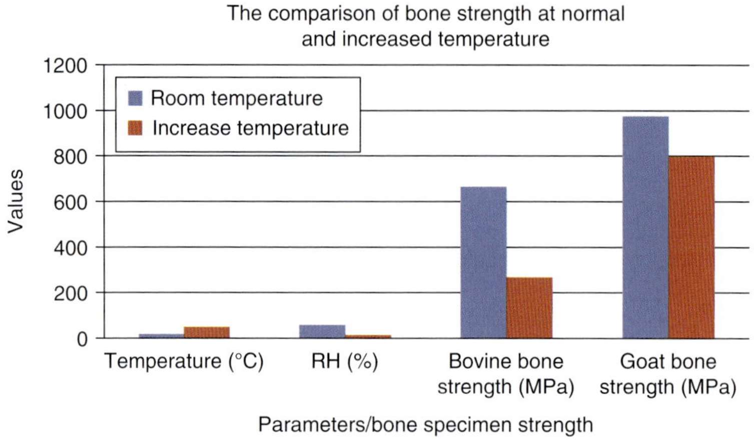


**Figure 1.14** Comparison of bone strength at normal and increased temperatures.

It is believed that the material properties of the goat bone tissue resulted in increased elastic modulus and a stiffer bone is formed. Comparison of the Young's modulus shows that goat bone (7.057, 82.49 GPa) has a higher strength than bovine bone (3.15, 82.22 GPa) at 22 and 55 °C respectively. However, at 85 °C the goat bone (4.616 GPa) has relatively lower Young's modulus than bovine bone (7.046 GPa). This is comparable to the results from Guocheng and Zreiqat [20].

There is a hypothesis that both bovine and goat bones aged at the highest temperature (85 °C) have the highest number of cracks than those aged at 22 and at 55 °C. The reason is that when exposed to high temperature, the bones experience higher mass loss and breakdown of hydroxyapatite. These losses result in cracks and fissures on the surface of the bones [33]. Table 1.6 shows the fracture appearance of different specimens.

**Table 1.6** Comparison of how the bone fracture in bovine and goat bone appears.

| Bovine specimen | Goat specimen |
|---|---|
| B1(C3) 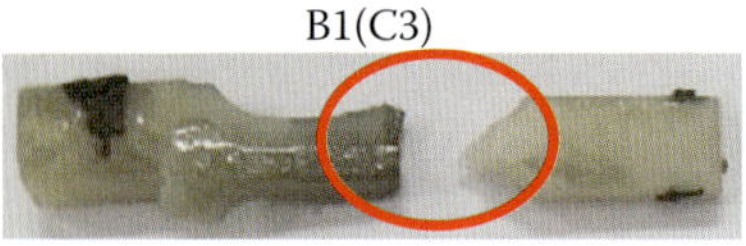 | G1  |
| B2(C4)  | G2  |
| B3(C2) 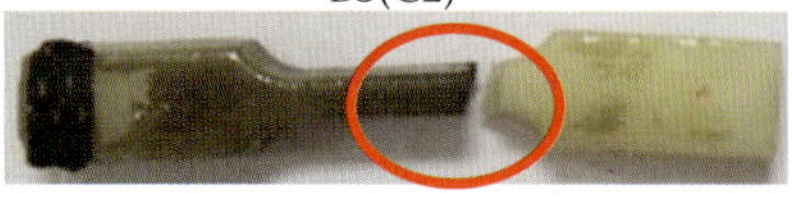 | G3  |

From Table 1.6, it can be seen that fractures happen at different parts of the bone, such as at the middle or the end of the gauge length. This is because cracks occur at the weak structural areas of bone due to tensile load when the specimen is pulled. Normally, the larger void spaces between organic and inorganic materials will be more susceptible to fracture. Besides, sliding occurs in undesirably oriented crystals before the tensile stress reaches the yield point. When loaded with tensile force, sliding happens along the crystalline plane of grains. Because the elastic properties of each crystal are different in different directions, the tensile stress may not be uniformly distributed [30, 32].

Moreover, the calcium contents in the bone itself can influence the fracture of bone. Calcium is an essential nutrient for strong bones. Calcium is needed to support the bone structure and for its hardness. When the calcium content is high, more calcium oxide is formed to form the bone. Hence, fractures will not occur easily. For example, the goat bone has higher calcium content (9.03 wt%) compared to bovine bone (5.67 wt%). So, it is harder for the goat bone to sustain a fracture regardless of the thickness compared to bovine bones.

Table 1.7 shows a comparison of the elastic modulus, tensile strength, and ductility between bovine and goat bones.

On comparing the bovine and goat mechano properties in Table 1.7, the test results show an almost consistent trend for each bone specimen. At 55 °C, both bones have the highest elastic modulus, highest ultimate tensile strength, and lowest ductility. At 85 °C, the bones have the lowest elastic modulus, lowest ultimate tensile strength, and medium ductility. Comparatively, the mechanical properties of goat bone are better than those of bovine bone. The elastic modulus of goat bone was higher than that of bovine bone. The high elastic modulus indicates that the goat bone has higher rigidity. It is a stiffer material and needs a high load to elastically deform.

Although the goat bone has higher modulus of elasticity and ultimate tensile strength, it exhibits brittleness behavior. Brittleness is the tendency of a material to fracture when there is no or little plastic deformation before failure. Figure 1.15 shows the stress–strain curve where curve "A" would be considered brittle, and curve "B" would be considered ductile [33]. By comparing this, it can be seen that the graph obtained from the experiment most likely follows curve "A" apart from having low ductility as observed in Figures 1.8–1.13.

**Table 1.7** Mechano properties of bovine and goat bones.

| Specimen | Temperature (°C) | Elastic modulus (MPa) | Tensile strength (MPa) |
|---|---|---|---|
| B3 | 20 | 3.15 | 54.27 |
| B4 | 50 | 82.22 | 91.37 |
| B2 | 80 | 0.6 | 0.41 |
| G3 | 22 | 7.06 | 128.25 |
| G4 | 55 | 82.49 | 129.64 |
| G2 | 85 | 0.78 | 1.20 |

B: Bovine, G: goat.

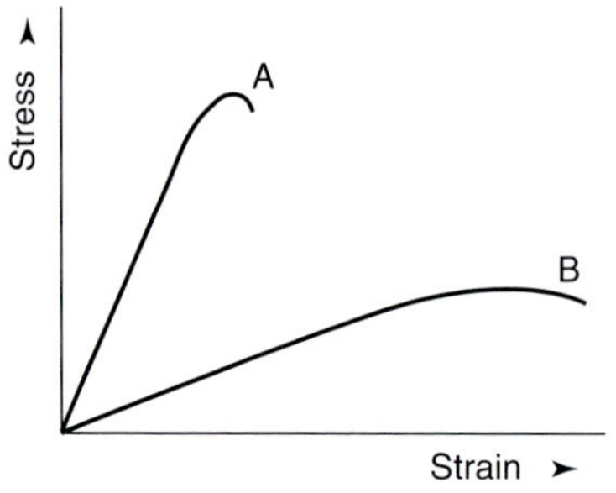


**Figure 1.15** Stress–strain curve. *Source:* Asmussen 2009 [33]. Reproduced with permission from Elsevier.

After conducting the tensile test, the hardness of both bovine and goat bone specimens were tested and analyzed under a micro/nano-indenter. From Figure 1.16, the general shapes of the loading and unloading curves are similar to the observation gained from human cortical and trabecular bones [26].

A summary of the nano-indentation results for both bovine and goat bones at different temperatures is presented in Table 1.8. The hardness, maximum depth of indentation, maximum load of indenter applied, and reduced modulus were recorded.

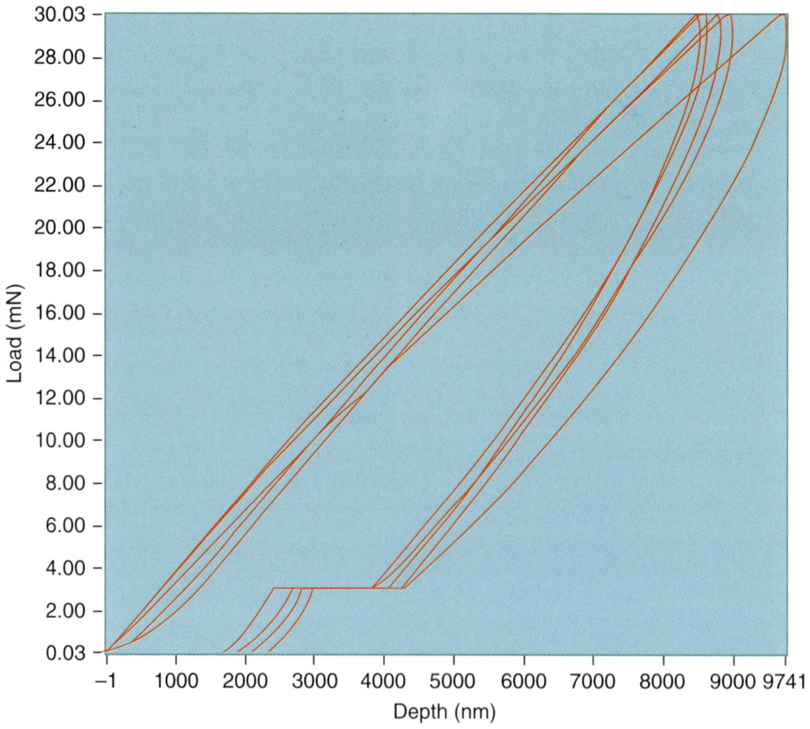


**Figure 1.16** Load versus indentation depth graph for bovine bone specimen.

**Table 1.8** Nano-indentation results for bovine and goat bones at different temperatures.

| Specimen | Hardness (GPa) | Max indentation depth (nm) | Maximum load (mN) | Reduced modulus (GPa) |
|---|---|---|---|---|
| B3(20) | 0.66663 | 1567.49 | 30.03 | 14.0437 |
| B4(50) | 0.18109 | 3636.29 | 30.03 | 1.46692 |
| B2(80) | 0.04077 | 8934.23 | 30.03 | 0.20847 |
| G3(20) | 0.24599 | 1523.72 | 8.03 | 3.63481 |
| G4(50) | 0.03662 | 7096.85 | 30.03 | 0.67908 |
| G2(80) | 0.29998 | 1308.03 | 8.03 | 4.34941 |

B: Bovine, G: goat.

Based on Table 1.8, the hardness of both bovine and goat specimens was almost similar to that of the femur bone in humans with a hardness of 0.55 ± 0.15 GPa [32]. In comparison, other implants such as metallic biomaterials including stainless steel and Co-based alloy had low surface hardness of about 38 ± 8.2 MPa, which is lower than that of human femur bone. From Figure 1.17, the average hardness of metallic implant is relatively higher than the lowest hardness of bone by about 3.77% as compared with data from Guocheng and Zreiqat [20].

In comparison with the normal bone with the highest average hardness, the marketed implants had lower hardness values of about 43% as shown in Figure 1.18. Hence, the bovine and goat bones still have higher hardness properties compared to metallic biomaterials. Moreover, these metallic implants had a high tendency for corrosion and wear in biological environments [20].

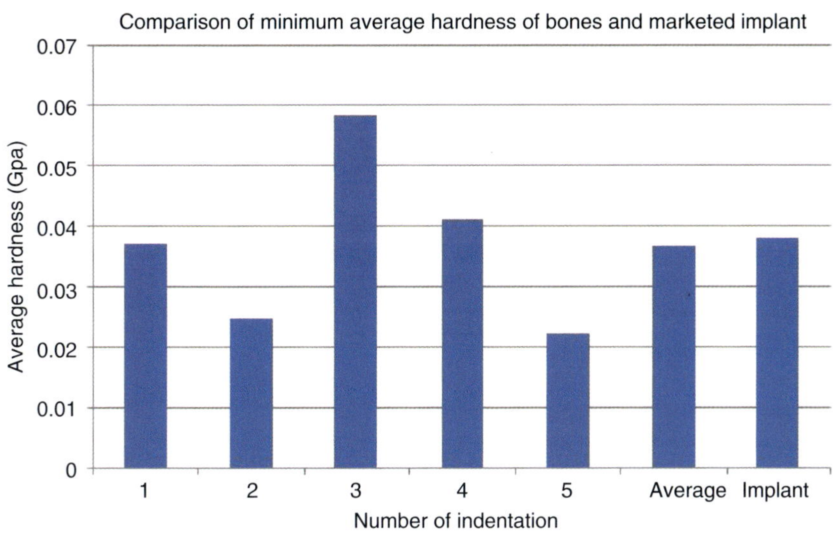


**Figure 1.17** Comparison of minimum average hardness of bone and implant.

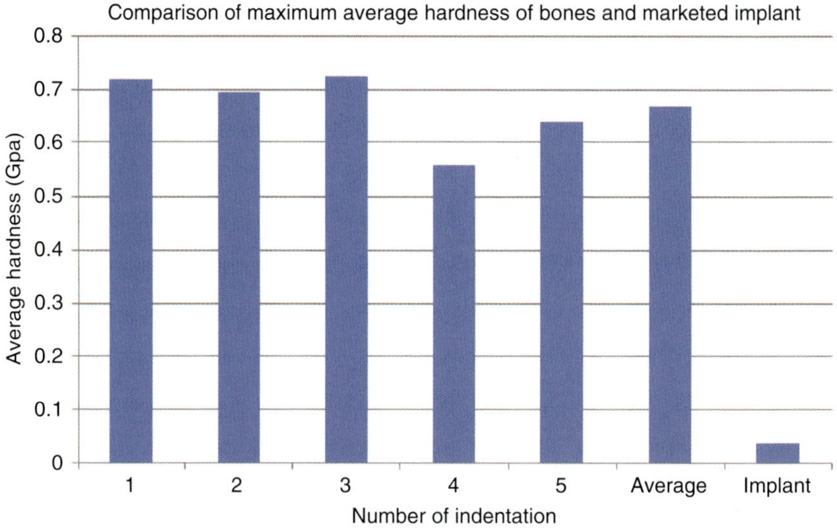


**Figure 1.18** Comparison of maximum average hardness of bone and implant.

The bone chosen for implant must be strong enough to withstand the mechanical loads. When it has better mechanical strength properties, it can sustain itself in the human body for long periods of time without undergoing repeat surgery. The weak mechanical properties of implant materials necessitate consistent replacement or repairing.

The present results and previous work done by researchers are slightly different. This may be due to the different anatomical location of animal body bones. In this experiment, the bone used originated from the femur of both bovine and goat models. But previously, most researchers had done the work on measuring the elastic modulus and hardness by using rib bone [23]. Moreover, it may due to different collagen orientation and distribution in bovine and goat bones. When the orientation and distribution are different, the tissue depth is different under the same load. This statement is also proved by ASCENZI that different distribution of osteon types can be differentiated by collagen fiber orientation and it can make the average properties distinctive [22].

From Table 1.8, one can observe the likely similarity to the result obtained by Rho. He also had shown that hardness is closely related to the Young's modulus [28]. It can be seen that when the hardness increases, Young's modulus also increases. In Malaysia, cow slaughtering in public is not much encouraged and cow bone is prohibited to be implanted due to religious issues. On the other hand, goat bone can be a better choice since it has relatively significant hardness properties and is widely accepted by every religion in Malaysia.

At the same time, it can be seen that both bovine and goat bone specimens have better hardness properties after adding a resin coating. A comparison of hardness with and without resin coatings is shown in Table 1.9.

**Table 1.9** Comparison between bone with resin and bone without resin.

| | Hardness (GPa) | |
|---|---|---|
| **Bone coating** | **Bovine** | **Goat** |
| Without resin | 0.024 91 | 0.069 73 |
| With resin | 0.666 63 | 0.245 99 |

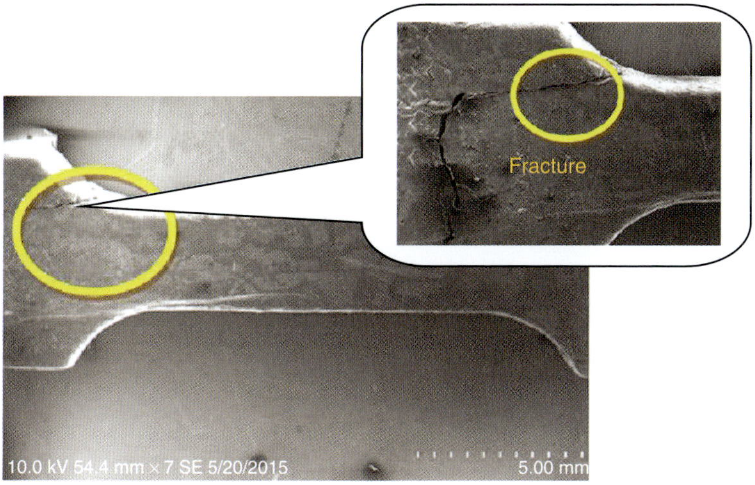


**Figure 1.19** The fracture structure of bone.

From Table 1.9, it is seen that the bovine and goat bone's hardness without resin hybrid is 0.024 91 and 0.069 73 GPa respectively. It increased to 0.666 63 and 0.245 99 GPa respectively after the bone was coated with resin hybrid. Epoxy resin is a thermoset that consists of single molecules of epoxyl resin and curing agents (hardeners). Both epoxyl resin and hardeners combine together chemically and form long chain of molecules [1]. When it is cured in the drying cabinet, it can become a hard polymer due to the formation of stable adhesive cross-linkages [24].

In order to investigate the microstructure of bones, the fracture surface specimens were observed by using SEM. Figure 1.19 shows the fracture structure for bovine bone specimen, which had broken at the edge of gauge length after tensile testing.

The SEM image in Figure 1.20 shows the abrasive elongation of bone while tensile loading is applied in the direction of elongation (yellow arrow) outward from the point where the voids propagation initiates and breaks at the end.

The abrasive wear elongation can be calculated by measuring the average distance between two parallel lines (yellow lines) as shown in Figure 1.21.

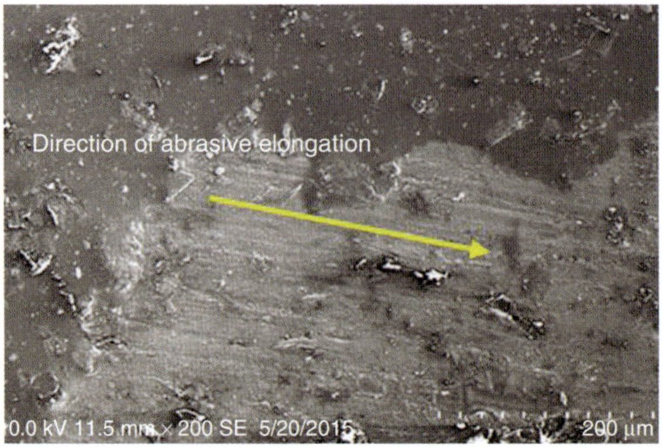


**Figure 1.20** Abrasive elongation of bone.

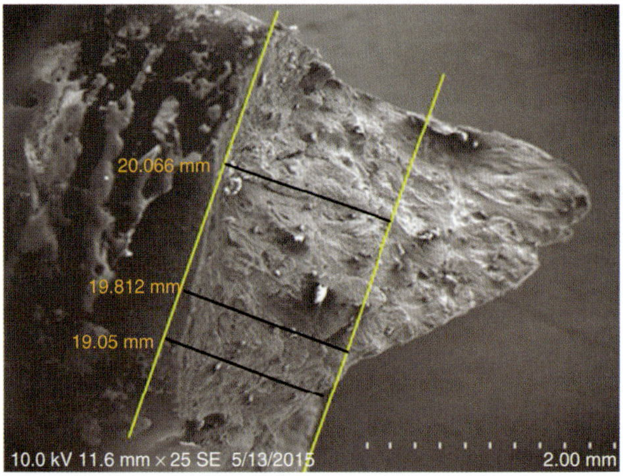


**Figure 1.21** The abrasive wear elongation of bone.

The average distance for the wear elongation before breaks of the bovine bone specimen (C3/B3) is 19.643 mm.

When the cross-section areas of bone were analyzed, it is noticeable that porosity exists in both bovine and goat bones as shown in Table 1.10.

On comparison of bovine and goat specimens as shown in Table 1.10, the goat bone has less porosity than bovine bone. The goat bone has smaller and narrower void spaces while the bovine bone has larger and wider void spaces. When monitoring the accelerated weathering at room temperature, the bovine bone has long-chain small interconnected porosity but the goat bone has more of small porosity. At 85 °C of weathering, the bovine bone has large volume porosity and goat bone has more of small size porosity. This has proved that when the porosity is high, the strength of the bone is less. The hypothesis that stated that both bovine and goat bones have the highest tensile strength is accepted.

**Table 1.10** Comparison of porosity in bovine and goat bones.

| | |
|---|---|
| 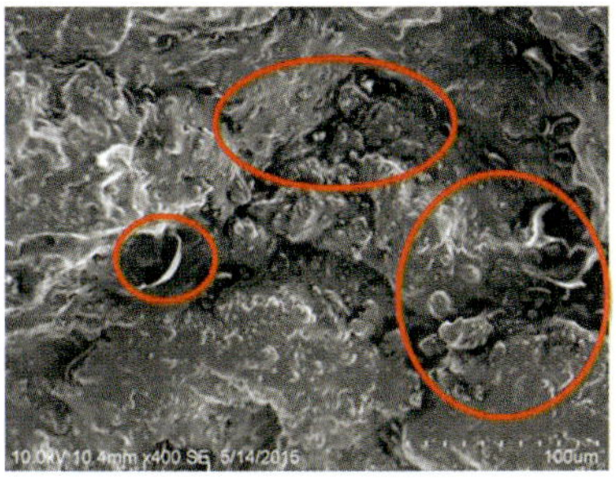 | 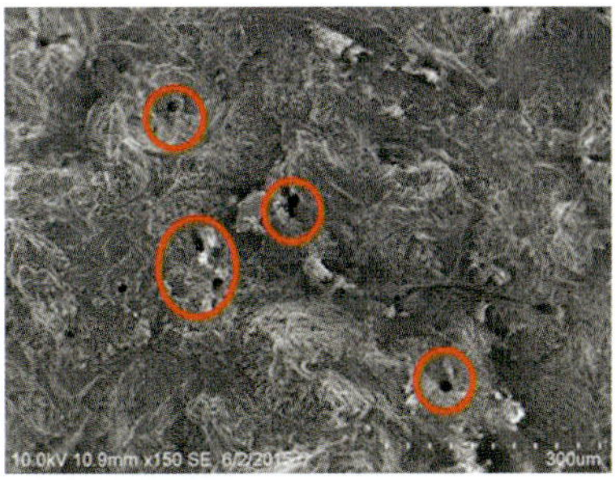 |
| Bovine bone at 22 °C (B1) | Goat bone at 22 °C (G1) |
| 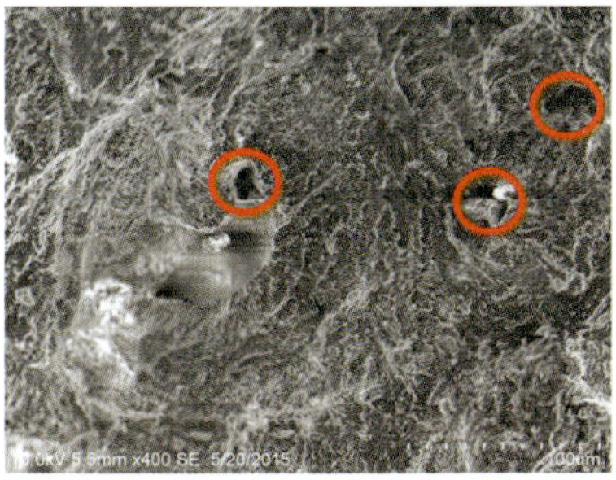 | 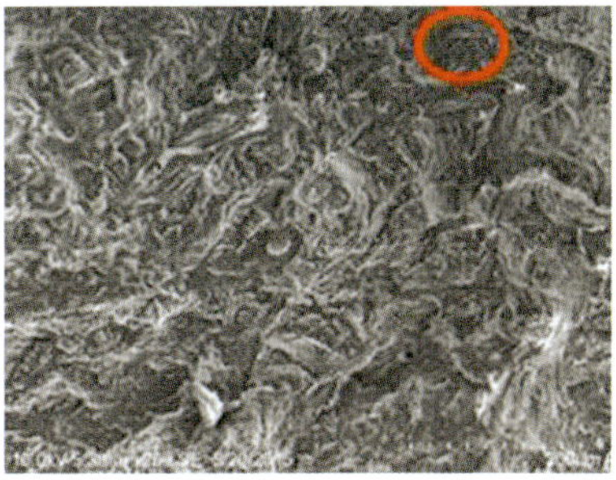 |
| Bovine bone at 55 °C (B2) | Goat bone at 55 °C (G2) |
| 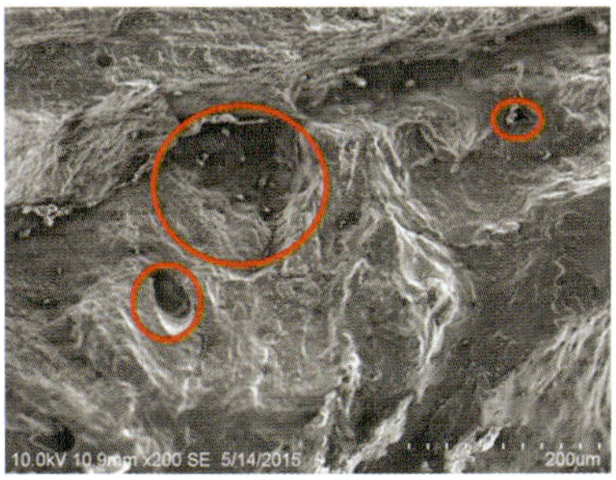 | 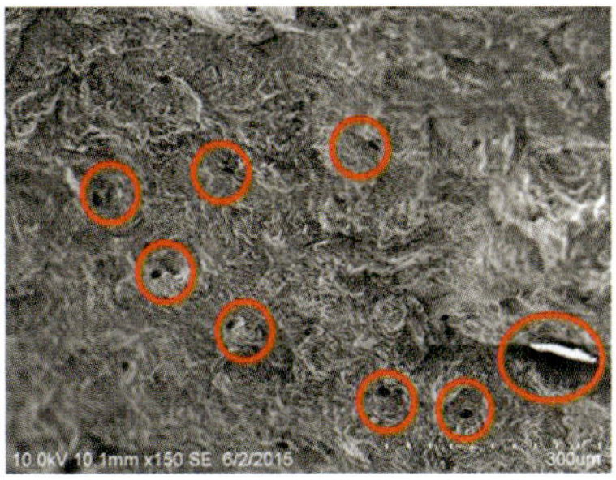 |
| Bovine bone at 85 °C (B3) | Goat bone at 55 °C (G3) |

Based on the results of EDAX analysis as shown in Figures 1.22–1.25, both bovine and goat bones were not pure organic biomaterials because they consist of other inorganic elements such as sodium, magnesium, and aluminum. From Figure 1.22 it can be seen that bovine bone consists mainly of carbon and hydrogen.

Besides, the EDAX analysis has shown that calcium content in goat bone is higher than that in bovine bone. Figures 1.22 and 1.23 show the overall calcium composition in both bovine and goat bone specimens prepared for testing at 22 °C (B1 and G1) and 85 °C (B3 and G3).

Figures 1.24 and 1.25 show that the goat bone has relatively higher calcium content compared to bovine bone. Higher composition of calcium can enhance the tensile strength of bone. So, the hypothesis that goat bone is stronger than the bovine bone is accepted.

| Element | Weight % | Atomic % | Net Int. | Net Int. Error |
|---|---|---|---|---|
| C K | 73.22 | 84.23 | 412 | 0.02 |
| O K | 12.61 | 11.06 | 41.57 | 0.1 |
| AlK | 0.4 | 0.21 | 10.55 | 0.16 |
| SiK | 0.39 | 0.19 | 10.55 | 0.16 |
| P K | 4.02 | 1.79 | 87.86 | 0.04 |
| NbL | 1.81 | 0.27 | 18.73 | 0.16 |
| PdL | 1.28 | 0.17 | 8.95 | 0.33 |
| CaK | 6.07 | 2.09 | 87.19 | 0.05 |

Bovine bone (B1)

| Element | Weight % | Atomic % | Net Int. | Net Int. Error |
|---|---|---|---|---|
| C K | 39.95 | 55.81 | 210.51 | 0.02 |
| O K | 28.31 | 29.69 | 153.44 | 0.02 |
| NaK | 0.55 | 0.4 | 15.7 | 0.11 |
| MgK | 0.33 | 0.23 | 11.81 | 0.16 |
| AlK | 0.34 | 0.21 | 12.2 | 0.15 |
| P K | 10.56 | 5.72 | 331.45 | 0.01 |
| HgM | 1.24 | 0.1 | 13.48 | 0.16 |
| CaK | 18.72 | 7.84 | 392.09 | 0.01 |

Goat bone (G1)

**Figure 1.22** Calcium composition in bone specimens at 22 °C (B1 and G1).

| Element | Weight % | Atomic % | Net Int. | Net Int. Error |
|---|---|---|---|---|
| C K | 73.71 | 85.99 | 419.44 | 0.02 |
| O K | 10.17 | 8.91 | 34.52 | 0.16 |
| NaK | 0.39 | 0.24 | 9.26 | 0.17 |
| Mgk | 0.36 | 0.21 | 9.97 | 0.2 |
| AlK | 0.46 | 0.24 | 12.63 | 0.18 |
| SiK | 0.36 | 0.18 | 10.08 | 0.2 |
| Y L | 0.35 | 0.06 | 3.95 | 0.82 |
| P K | 3.66 | 1.66 | 81.84 | 0.05 |
| NbL | 1.62 | 0.24 | 17.16 | 0.18 |
| ClK | 0.18 | 0.07 | 3.69 | 0.68 |
| RaM | 1.06 | 0.07 | 8.01 | 0.35 |
| CaK | 5.67 | 1.98 | 84.79 | 0.05 |
| HfL | 1.99 | 0.16 | 1.36 | 0.71 |

Bovine bone (B3)

| Element | Weight % | Atomic % | Net Int. | Net Int. Error |
|---|---|---|---|---|
| C K | 58.39 | 70.54 | 395.97 | 0.01 |
| O K | 25.34 | 22.98 | 126.27 | 0.03 |
| NaK | 0.38 | 0.24 | 10.47 | 0.15 |
| MgK | 0.37 | 0.22 | 12.52 | 0.13 |
| P K | 5.63 | 2.64 | 162.39 | 0.02 |
| PdL | 0.85 | 0.12 | 8.09 | 0.33 |
| CaK | 9.03 | 3.27 | 173.25 | 0.02 |

Goat bone (G3)

**Figure 1.23** Calcium composition in bone specimens at 85 °C (B1 and G1).

EDS1 kV: 15 Mag:200 Takeoff:35.1 Live time:30 Amp time:6.4 Resolution:125.5

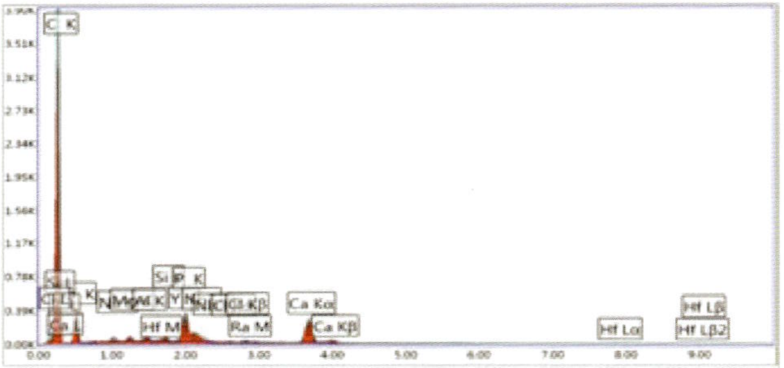


| Element | Weight % | Atomic % | Net Int. | Net Int. Error |
|---|---|---|---|---|
| C K | 73.71 | 85.99 | 419.44 | 0.02 |
| O K | 10.17 | 8.91 | 34.52 | 0.16 |
| NaK | 0.39 | 0.24 | 9.26 | 0.17 |
| MgK | 0.36 | 0.21 | 9.97 | 0.2 |
| AlK | 0.46 | 0.24 | 12.63 | 0.18 |
| SiK | 0.36 | 0.18 | 10.08 | 0.2 |
| Y L | 0.35 | 0.06 | 3.95 | 0.82 |
| P K | 3.66 | 1.66 | 81.84 | 0.05 |
| NbL | 1.62 | 0.24 | 17.16 | 0.18 |
| ClK | 0.18 | 0.07 | 3.69 | 0.68 |
| RaM | 1.06 | 0.07 | 8.01 | 0.35 |
| CaK | 5.67 | 1.98 | 84.79 | 0.05 |
| HfL | 1.99 | 0.16 | 1.36 | 0.71 |

**Figure 1.24** EDAX analysis for bovine specimen for testing at 85 °C (B3).

| EDS1 | kV: 15 | Mag:200 | Takeoff:35 | Live time:30 | Amp time:6.4 | Resolution:125.5 |
|---|---|---|---|---|---|---|

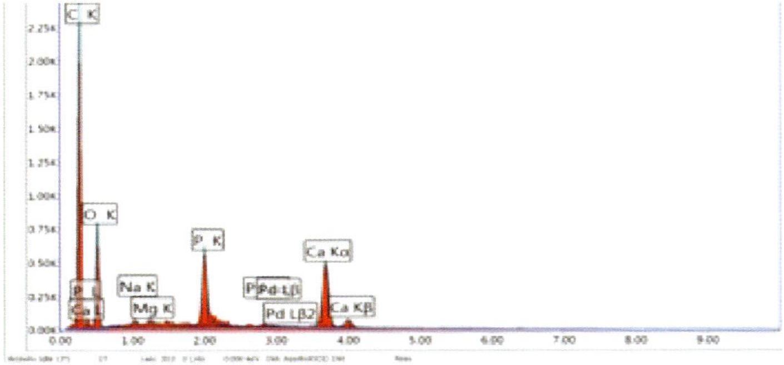


| Element | Weight % | Atomic % | Net Int. | Net Int. Error |
|---|---|---|---|---|
| C K | 58.39 | 70.54 | 395.97 | 0.01 |
| O K | 25.34 | 22.98 | 126.27 | 0.03 |
| NaK | 0.38 | 0.24 | 10.47 | 0.15 |
| MgK | 0.37 | 0.22 | 12.52 | 0.13 |
| P K | 5.63 | 2.64 | 162.39 | 0.02 |
| PdL | 0.85 | 0.12 | 8.09 | 0.33 |
| CaK | 9.03 | 3.27 | 173.25 | 0.02 |

**Figure 1.25** EDAX analysis for goat specimen for testing at 85 °C (G3).

## 1.5 Conclusion

Exposure of animal bones to accelerated aging condition changes their mechanical properties. When a bone is exposed to high temperature environment for a long time, the mass loss of the bone is higher. Based on the results obtained, it can be concluded that the highest tensile strength (958.18 N for bovine bone and 1031.88 N for goat bone) was obtained at moderate temperature (around 55 °C) and moderate RH (15.08% RH). At low temperatures such as room temperature, the bone has lower tensile strength due to greater susceptibility to mold growth in high humidity environment. At very high temperatures, the bone deteriorates and cracks easily due to an over-drying environment. Hence, the lowest tensile strength is obtained for bovine (501.64 N) and goat bones (713.06 N) at 85 °C.

Besides, the goat bone is a better choice with good mechanical properties because it has higher tensile strength compared to bovine bone. The reason is that the physical activity in goat is more frequent than in the cow and hence its collagen network in bone is enhanced. Calcium element content in goat bone during EDAX analysis is higher than that in bovine bone. From the micrograph obtained from SEM analysis, it is seen that the porosity in goat bone is lesser, smaller, and narrower than that in bovine bone. All of these results prove that goat bone has better tensile strength compared to bovine bone.

In hardness test, both bovine and goat bones have shown good hardness properties, which are nearly similar to the human bone's hardness (0.55 ± 0.15 GPa). They have shown higher hardness (ranging from 0.181 09 to 0.6663 GPa) compared to the existing marketed metallic implants (38 ± 8.2 MPa). However, bovine bone biomaterials must be carefully selected to avoid any sensitive religious issues, in Malaysia particularly. For this reason, the goat model may still be considered as a good option.

## 1.A Relative Humidity Chart

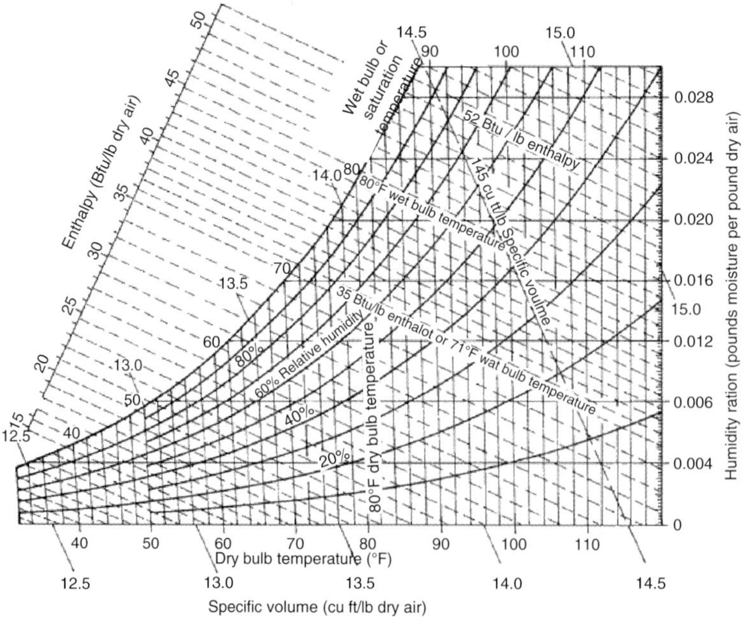

## References


1 A. Wadham. "Module 1-basic material properties and how they can be altered". Canteach, 2 June 1993. Available from: https://canteach.candu.org/Content%20Library/20040901.pdf (accessed 23 June 2015).

2 Qwamizadeh, M., Zhang, Z., Zhou, K., and Zhang, Y.W. (2015). On the relationship between the dynamic behavior and nano-scale staggered structure of the bone. *Journal of the Machanics and Physics of Solids* 78: 17–31.

3 Pilitsis, J.G., Lucas, D.R., and Rengachary, S.R. (2002). Bone healing and spinal fusion. *Neurosurgical Focus* 13 (6), e1. PMID 15766227.

4 Isaksson, H., Tolvanen, V., Finnilä, M.A.J. et al. (2009). Long-term voluntary exercise of male mice induces more beneficial effects on cancellous and cortical bone than on the collagen matrix. *Experimental Gerontology* 44 (11): 708–717.

5 Hiller, M.L. and Bell, L.S. (2007). Differentiating human bone from animal bone: a review of histological methods. *Journal of Forensic Sciences* 52 (2): 249–263.

6 Pearce, A.I., Richards, R.G., Milz, S. et al. (2007). Animal models for implant biomaterial research in bone: a review. *European Cells and Materials* 13: 1–10.

7 Prothero, D.R., Manning, E.M., and Fischer, M.S. (1988). The phylogeny of the ungulates. In: *The Phylogeny and Classification of the Tetrapods*, vol. 2 (ed. M.J. Benton), 201–234. Oxford: Oxford University Press.

8 ASTM (1997). *E8 Standard Test Methods of Tension Testing of Metallic Materials*. s.l.:s.n.

9 ASTM (1997). Standards methods of testing small clear specimens of timber. *Annual Book of ASTM Standards Des. ASTM D 143-94*, vol. 4.10.

10 D. Subit, E.D.P. de Dios, J. Valazquez-Ameijide, C. Arregui-Dalmases, J. Crandall, 2013. Tensile material properties of human rib cortical bone under quasi-static and dynamic failure loading and influence of the bone microstructure on failure characteristics. Biological Physics, v2, 1–22. Available from: http://arxiv.org/pdf/1108.0390v2.pdf.

11 Ajadi, D.A. and Sanusi, Y.K. (2013). Effect of relative humidity on oven temperature of locally design solar carbinet dryer. *Global Journal of Science Frontier Research Physics and Space Science* 13 (1).

12 Huynh, T.T., Aarnink, A.J., Verstegen, M.W. et al. (2005). Effects of increasing temperatures on physiological changes in pigs at different relative humidities. *Journal of Animal Science* 83 (6): 1385–1396.

13 Karr, L.P. and Outram, A.K. (2012). Bone degradation and environment: understanding, assessing and conducting archaeological experiment using modern animal bones. *International Journal of Osteoarcgaeology* 25: 201–212.

14 Magargee, J., Cao, J., Zhou, R. et al. (2011). Characterization of tensile and compressive of microscale sheet metals using a transparent microwedge device. *Journal of Manufacturing Science and Engineering* 133 (6): 064501.

15 Zhang, W., Tekalur, S.A., Baumann, M., and McCabe, L.R. (2013). The effects of damage accumulation on the tensile strength. *Journal of Biomechanics* 46 (2013): 964–972.

16 Feng, L. and Jasiuk, I. (2010). Effect of specimen geometry on tensile strength of cortical bone. *Journal of Biomedical Materials Research Part A* 95 (2): 580–587. doi: 10.1002/jbm.a.32837.
17 Warhatkar, H., Chawla, A., Mukherjee, S., and Malhotra, R. (2008). Experimental study of variation between quasi-static and dynamic load deformation properties of vovine medial collateral ligaments. SAE International, 09B-0276 also in Biomechanics, 2009 – SP-2225.
18 Schaffler, M.B. and Burr, D.B. (1988). Stiffness of compact bone: effects of porosity and density. *Journal of Biomechanics* 21 (1): 13–16.
19 Kohan, N., Via, B.K., and Taylor, S. (2011). A comparison of geometry effect on tensile testing of wood strands. *Forest Products Journal* 62 (3): 167–170.
20 Guocheng, W. and Zreiqat, H. (2010). Functional coatings or films for hard-tissue applications. *Materials* 3: 3994–4050. doi: 10.3390/ma3073994.
21 Reilly, D.T. and Burstein, A.H. (1975). The elastic and ultimate properties of compact bone tissue. *Journal of Biomechanics* 8 (6): 393–405.
22 Lau, M.L., Lau, K.T., Yeo, Y.D.Y. et al. (2010). Measurement of bovine bone properties through surface indentation technique. *Materials and Manufacturing Processes* 25 (5): 324–328.
23 Lau, M.L., Lau, K.T., Ku, H. et al. (2012). Measurement of heat treatment effects on bovine cortical bones by nanoindentation and compression testing. *Journal of Biomaterials and Nanobiotechnology* 3 (1): 105–113.
24 Tiwari, A. (2012). Nanomechanical analysis of hybrid silicones and hybrid epoxy coatings – a brief review. *Advances in Chemical Engineering and Science* 2 (1): 34–44.
25 Bensamoun, S., Fan, Z., Brice, I. et al. (2008). Assessment of mechanical properties of human osteon lamellae exhibiting various degrees of mineralization by nanoindentation. *Journal of Musculoskeletal Resarch.* 11 (3): 135–143.
26 Rho, J.Y., Tsui, T.Y., and Pharr, G.M. (1997). Elastic properties of human cortical and trabecular lamellar bone measured by nanoindentation. *Biomaterials* 18 (20): 1325–1330.
27 Rho, J.Y., Zioupos, P., Currey, J.D., and Pharr, G.M. (2002). Microstructural elasticity and regional heterogenity in human femoral bone of various ages examined by nano-indentation. *Journal of Biomechanics* 35: 189–198.
28 Rho, J.Y. and Pharr, G.M. (1999). Effect of drying on the mechanical properties of bovine femur measured by nanoindentation. *Journal of Materials Science Materials in Medicine* 10 (8): 485–488.
29 Ubelaker, D.H. (2009). The forensic evaluation of burned skeletal remains: a synthesis. *Forensic Science International* 183 (2009): 1–5.
30 Stiner, M.C., Kuhn, S.L., Weiner, S., and Bar-Yosef, O. (1995). Differential burning, recrystallization, and fragmentation of archaeological bone. *Journal of Archaeological Science* 22: 223–237.
31 Anthony Wilkes, D.W. (2015). Measurement of humidity. *Anaesthesia and Intensive Care Medicine* 16 (3): 128–131.

32 Heritage, C. (2015). *Conservation of Wet Faunal Remains: Bone, Antler, and Ivory*. Canada: Canada Conservation Institutes Notes 4/3. Available: https://www.canada.ca/content/dam/cci-icc/documents/services/conservation-preservation-publications/canadian-conservation-institute-notes/4-3-eng.pdf?WT.contentAuthority=4.4.10 (accessed 20 March 2017).

33 Asmussen, B. (2009). Intentional or incidental thermal modification? Analysing site occupation via burned bone. *Journal of Archaeological Science* 36 (2): 528–536.

# 2

# Nanofracture and Wear Testing on Natural Bones

*Ramdziah M. Nasir, Muhammad A.A. Muhmad, and Abdul Y. Saad*

*University of Science, Malaysia, School of Mechanical Engineering, Engineering Campus, Nibong Tebal 13400, Penang, Malaysia*

## 2.1 Introduction

Bones form an important part of the endoskeleton of vertebrates. They play a role in the support and protection of the various organs of the body. Furthermore, bones also provide a frame to keep the body supported and a leverage system for body movement. Bones come in several shapes and although they have a complex, lightweight, strong, and hard structure, they are vulnerable to impact, which leads to fractures internally and externally. Fractures normally occur near joints and are common among people with rather excessive activities such as sportspersons, astronauts, and hard-working laborers. Whenever someone slips and is about to fall, there is a tendency that he/she will detain his/her posture before falling to the floor or ground. This will have an impact on the joint, causing fracture.

Fractures can be classified into two categories, simple and compound fractures. Simple fractures involve broken bones that remain within the body and do not penetrate the skin whereas compound fractures involve broken bones that penetrate through the skin and expose the bone and deep tissues to the external environment as shown in Figure 2.1. This type of fracture will lead to infection.

Nowadays, in medical application synthetic bones are used to repair fracture, and this technique is called bone grafting. Bone basically has the ability to grow completely but requires a very small fracture space or some kind of scaffold to do so. Bone grafts may be autologous, allograft, or synthetic with similar mechanical properties to bone. Most bone grafts are expected to be reabsorbed and replaced as the natural bone heals over a few months' time. They normally guide the natural bone to regenerate [2].

Bone has a complex structure that consists of both cellular and extracellular matrices composed of dense crystalline fibers and minerals such as calcium phosphate, calcium carbonate, and others that are responsible for rigidity, porosity, and hardness to resist the compression that is caused by force or impact [3].

*Synthesis and Tribological Applications of Hybrid Materials*, First Edition.
Edited by Mohammad Jawaid, Rajini Nagarajan, Jacob Sukumaran, and Patrick De Baets.

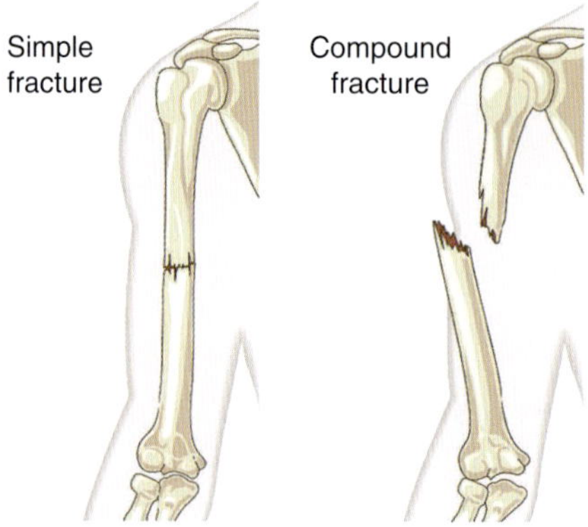


**Figure 2.1** Types of fracture. *Source:* Encyclopædia Britannica 2006 [1]. Reproduced with permission from Britannica.

Furthermore, bones have three functions [4] in the body:

1) Bone is a major organ for calcium homeostasis and is an essential store of phosphate, magnesium, potassium, and bicarbonate.
2) Bone provides mechanical support for soft tissue and leverage for muscle action.
3) Lastly, it is a major site of hematopoiesis in human adults.

In vertebrae, there are two types of bones, cortical bone (compact bone) and trabecular or cancellous bone (spongy bone). Compact bone is the hard section that makes up the long cylindrical bone whereby osteon which contain hard calcified matrix bone cell lodged in space of concentric lamellae. Spongy bone is an interior trabeculae meshwork with thin, irregular shaped plate arranged in a latticework network. Also, bony tissue surrounds blood in the compact bone, while blood surrounds the bony tissue in the cancellous bone as shown in Figure 2.2 by Qin [5].

Low impact fracture (fragility fracture) is a type of pathologic fracture that occurs as a result of normal activities, such as a fall from standing height or less. Three fracture sites are said to be typical of fragility fractures: vertebral fractures, fractures of the neck of the femur, and Colles fracture of the wrist. This definition arises because a normal human being ought to be able to fall from standing height without breaking any bones, and a fracture therefore suggests weakness of the skeleton.

Fragility fracture is defined by the World Health Organization as "a fracture caused by injury that would be insufficient to fracture a normal bone, the result of reduced compressive and/or torsional strength of bone." Clinically, a fragility fracture may be defined as a fracture "...that occurs as a result of a minimal trauma, such as a fall from a standing height or less, or no identifiable trauma." Typical fracture sites in patients with osteoporosis include vertebra (spine), proximal femur (hip), distal forearm (wrist), and proximal humerus [6].

Wear is a process where two surfaces are sliding or bounding against each other, which results in dimensional loss of the material. There are three stages of wear: (a) early run-in period where the surfaces adapt to each other and the wear rate may vary from high to low; (b) mid-age process with a steady rate of wear;

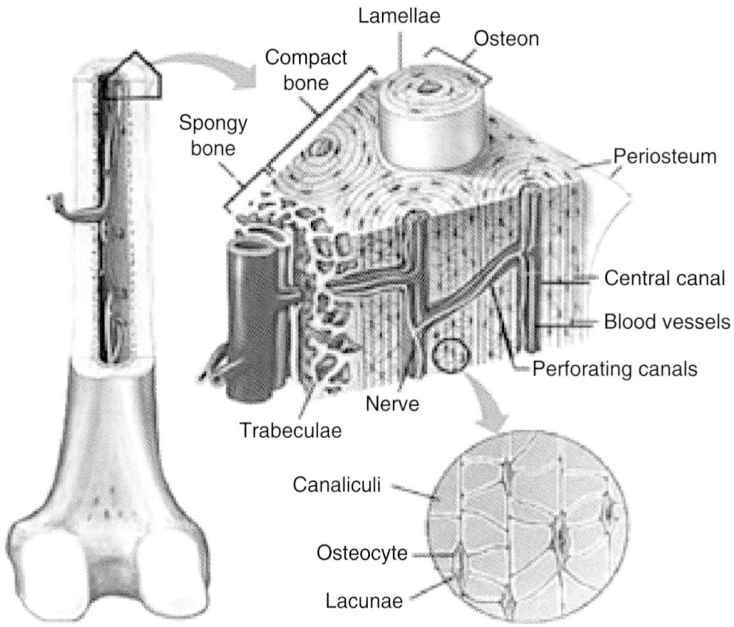


**Figure 2.2** The bone structure. *Source:* Qin 2013 [5]. Reproduced with permission from CRC Press.

and lastly (c) old-age period where the component give high aging rate and subject to rapid failure. The study of the processes of wear is part of the discipline of tribology. The complex nature of wear has delayed its investigation and has resulted in isolated studies toward specific wear mechanisms or processes [7].

There are several processes or mechanisms related to wear – adhesive wear, abrasive wear, surface fatigue, fretting wear, and erosive wear. However, the mechanism that needs to be emphasized is abrasive wear. Abrasive wear essentially occurs in the presence of hard rough friction against a softer surface. Various abrasive wear mechanisms causing surface destruction such as cutting, scratching, smearing, or repeated plastic deformation [8] have been reported.

There are two types of abrasive wear, two body wear and three body wear, as shown in Figure 2.3. Two body wear occurs when there are only two rubbing parts involved in the friction process of the wear whereas three body wear occurs when a hard particle (grit) is trapped between the rubbing surfaces.

McKellop et al. [9] set a goal for studies employing pin-on-disk testing to obtain a ranking of materials that is free from uncontrolled implant design effects, based on their relative wear rates rather than on absolute wear rates. Besides, Yu et al. [10] studied the fretting behavior of cortical bone against titanium and its alloy. Abrasive wear and delamination (micro-cracks) characterize the wear mechanism of cortical bones. Wang and Ge [11] reported using swine femoral bone against UHMWPE with dry friction, physiological water, and human plasma lubrication. In summary, the wear mechanism upon dry friction for compact bone is fatigue wear; under physiological water lubrication, it is corrosive wear; and with human plasma lubrication, it is abrasive wear. Zhu

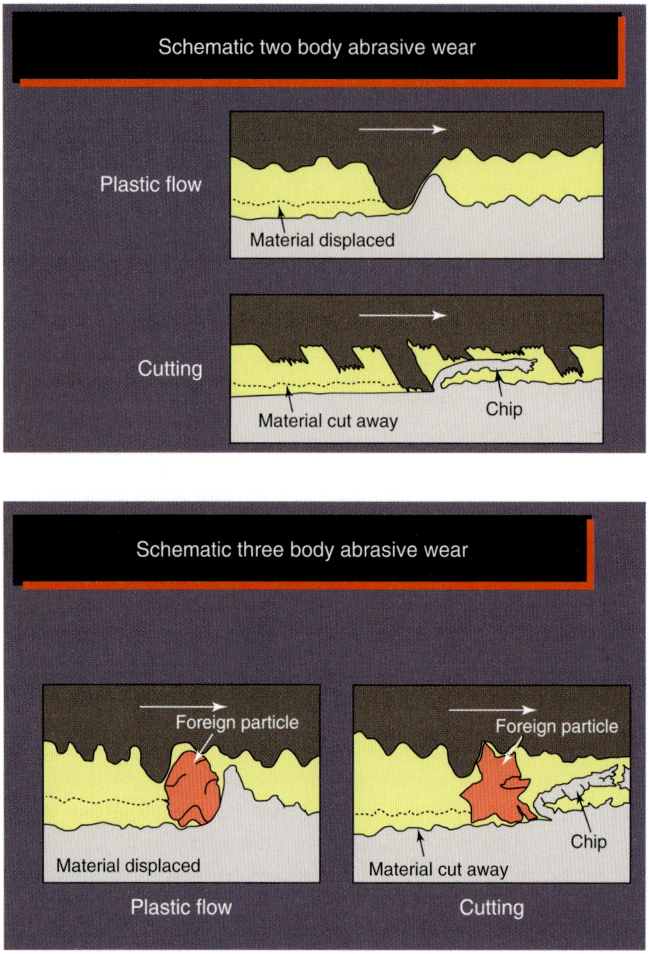


(a) (b)

**Figure 2.3** (a) Two body wear (b) three body wear. *Source:* McKellop et al. 1981 [9]. Reproduced with permission from John Wiley & Sons.

et al. [12] also have investigated the abrasion on natural bovine bone lubricated with water. Statistically, the impacts of interactions between the elected bivariants are insignificant on the wear of cancellous bone except for the sliding distance.

Now that clinically relevant wear mechanisms could be facilitated on multidirectional pin-on-disk testers, test parameters that would produce clinically relevant rankings of materials based on their wear rates had to be identified. Before proceeding with the summary of relevant test parameters, it should be noted that since the 1970s, the wear behavior of polyethylene has been reported as a wear factor. According to Archard's law (1953), the volume of wear debris generated is proportional to load and sliding distance by a wear factor, which is assumed to be constant, independent of the magnitude of the load. Another alternative parameter for reporting wear behavior is the volumetric/gravimetric wear per sliding distance/number of cycles, which does not require assumptions about the wear behavior. Wear factor, in theory, permits comparison of wear rates obtained with any loads and any sliding distances per cycle whereas wear volume per million cycles requires that the same testing conditions such as contact area and contact pressure be used if results from different studies were to be compared. Historically, wear rates were also converted to depth of wear (mm $year^{-1}$) by dividing volumetric wear by (nominal) contact area to permit comparisons with clinical radiographs [9]. Many researchers reported that wear resistance is inversely proportional to wear rate as in previous findings [13–15].

Tensile test is the most fundamental, simple, and fully standardized test on any material. By pulling the sample, one can determine how the material will elongate in response to the applied tension force, by obeying or denying Hooke's law, as shown in Figure 2.4.

Bowman et al. [16] had found that the results of stress–strain curves were similar to the mechanical tests of other collagenous tissues with an average modulus of 613 MPa (S.D. = 113 MPa), ultimate stress of 61.5 MPa (S.D. = 13.1 MPa), and ultimate strain of 12.3% (S.D. = 0.5%), respectively. Zhang et al. [17] observed that the average Young's modulus, $\epsilon$, of calleous bone is 21.5 GPa with an ultimate stress UT of 93.5 MPa longitudinally while $\epsilon$ is 10.9 GPa with UT of 36.2 MPa transversely. According to An and Draughn [18], the decision to use this parameter to conduct tensile test was based on ASTM 565. The speed of testing may be defined in terms of free running of the crosshead speed or in terms of rate of stressing the specimen. Both methods are permissible. When free-running crosshead speed is used, the speed shall be 0.50 10% mm (0.020 10% in.)/min. When rate of stressing the specimen is used, an applied force of 890 220 N (200 50 lbf)/min shall be used. Bowman et al. [16] stated that all the mechanical tests were performed under displacement control at a rate of 0.0025 mm $s^{-1}$ with a 100 lb load cell. The loading history used a 1 N preload and involved nine preconditioning cycles between ~0.0% and 5.0% strain before the final ramp to failure.

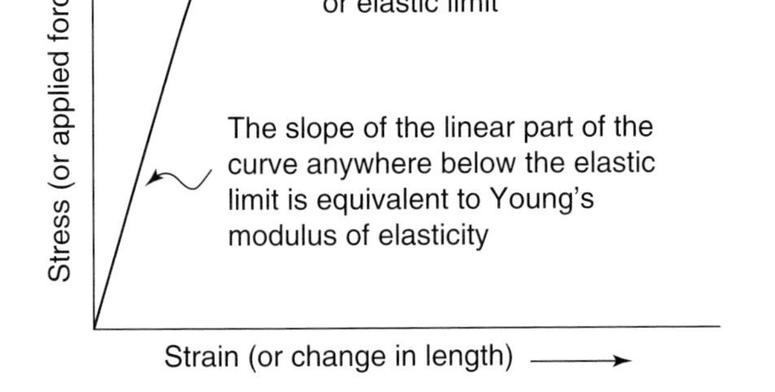


**Figure 2.4** Hooke's law.

Materials hardness not only restricted the capability to resist plastic deformation but also the resistance toward scratching, smearing, abrasion, cutting, fretting, bending, and penetrating resistance. Zysset et al. [19] used a nano-indentation technique with a custom irrigation system for simultaneously measuring the force and displacement of a diamond tip pressed 500 nm into moist bone tissue. An isotropic elastic modulus was calculated from the unloading curve with an assumed Poisson ratio of 0.3, while hardness was defined as the maximal force divided by the corresponding contact area. The elastic moduli ranged from 6.9 ± 4.3 GPa in trabecular tissue from the femoral neck of a 74-year-old female up to 25.0 ± 4.3 GPa in the interstitial tissue from the diaphyseal cortex of a 69-year-old female.

Furthermore, Dall'Ara et al. [20] found that the interstitial bone (13.2 ± 4.4 GPa and 44.7 ± 20.3 HV) was the stiffest and hardest compared to the osteonal bone (10.9 ± 3.8 GPa and 37.8 ± 17.3 HV) and the trabecular bone (8.1 ± 3.0 GPa and 28.8 ± 11.2 HV) transversely. In the axial direction, intact trabecular bone (11.4 ± 4.3 GPa) was 16% less stiff than the intact interstitial bone but comparable to the intact osteonal bone.

In this study, the use of natural bone as an alternative to synthetic materials such as polymer and plastic that are being used in surgery is examined. When synthetic material is used, it should be replaced in 5 years because it will go through excessive wear and cause debris production. This debris enters the blood circulation and causes infection. Thus, the use of natural bone is expected to be more beneficial compared to the synthetic material.

Research will be conducted based on several experiments to achieve the objectives of this project. The objectives of this project are to assess low impact fracture of natural bone focusing on wear, friction, and dependent parameters (applied normal load, sliding speed, experiment running duration, and the radius of the rotating effect). In this experiment, only cow's bone is used as the specimen to be tested because its thickness fulfills the dimensional requirements of the specimen. For the wear test, pin-on-disk tribometer (PODT) is used. Tensile test, hardness test, microstructure, and elemental composition using scanning electron microscope (SEM) imaging and energy dispersive analysis X-ray (EDAX) analysis are carried out to gain additional information about the specimen's properties.

## 2.2 Methodology

The methodology focuses on investigating low impact fracture of natural bones focusing on the effects of wear, friction, and various dependent parameters. This study is aimed at using natural bone as an alternative to synthetic bone usage in medical practice, and therefore, a comparative study of bone properties is necessary. The wear test is done using PODT. Experimental study is required to characterize the critical parameters of bones in terms of stability period and visualization. The natural bone used is cow bone, which is available from any wet market. The first step in the preparation of the specimen is the cooking process. The process of maceration can be speeded up via cooking. Boiling the bones can destroy disease-causing organisms and also facilitate the cleaning of unwanted substances such as flesh, bone marrow, and fat. The entire process takes from a few minutes to several hours depending on the size of the bones. The cooked meat is fairly easy to remove with a knife, scalpel, or pliers. Then, the bone is dried under sunlight (~48 h of sun) to remove the odor of bone and to prevent infection from bacteria that causes the bone to decay. Then, the bone is kept in desiccators to maintain it dry.

After the cleaning process, the second step is to make the bone smaller in size (about 10 mm of thickness) by cutting. The smaller precut bone is inserted into the spindle of a lathe machine. The center of the bone is located and locked by a tailstock quill. The purpose of the tailstock quill is to lock the specimen from swinging so that a symmetrical shape of the specimen can be produced. Then the water jet cutting tool is fed with the bone specimen by 1 mm inward until the desired diameter for the testing, 6 mm, has been achieved. The specimens are cleaned to remove dirt and contaminant substances present on them. The specimens are first immersed into an ultrasonic cleaner for about 3–5 min and then dried using the specimen blower.

The difference between initial and final mass of the specimens will represent its volume loss as in Equation (2.1).

$$\text{Volume loss}, \Delta V = \frac{\text{Initial mass} - \text{Final mass}, \Delta m}{\text{Density}, \rho} \tag{2.1}$$

Hence, the specimens are weighted before and after each test run by using a weighing balance to account for the material removal rate (wear rate) by using Equation (2.2).

$$\text{Material removal rate}, \text{mrr} = \frac{\text{Volume loss}, \Delta V}{\text{Sliding distance}, d} \tag{2.2}$$

This calculation will later be used to validate the resulting wear rate obtained from the experiment. Archimedes principle is applied to determine the specimens' volume as their shapes are irregular and complicated to be calculated theoretically.

This test was carried out by using PODT Tester Machine (Model No TR-20) that is available in the Nanotechnology Laboratory at the School of Mechanical Engineering, USM, Malaysia. The wear test was done to determine the wear rate of the specimens. A PODT consists of a stationary pin under an applied load in contact with a rotating disc as shown in Figure 2.5. The pin has many shapes to simulate a specific contact. The coefficient of friction (COF) is determined as the ratio of the frictional force to the loading force on the pin. Parameters such as applied load, radius of rotation, sliding speed, and running time are adjusted properly prior to usage.

First, the sliding table was adjusted in order to set the radius of rotation. Then, the screws were tightened to prevent the table from sliding. After that, the stopper was locked using the level mechanism and the specimen was inserted into the specimen holder and firmly tightened. Then, the stopper was released and the desired load mounted on the machine. After that, the PODT

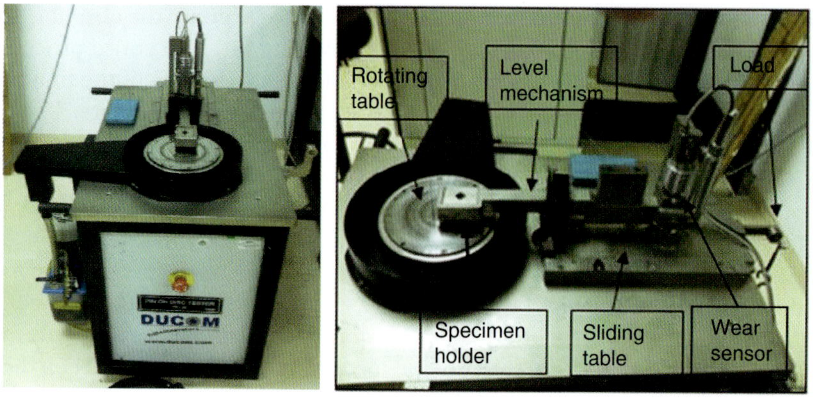


**Figure 2.5** Pin-on-disk tribometer setup.

was set up. The power source and PC that was linked with the tester were switched on. The wear sensor was checked for proper functioning by lifting away the probe from the surface. The wear rate reading should be around 3500 $mm^3 N^{-1} m^{-1}$, indicating that the sensor is functioning well. The level of the sensor was adjusted to get zero reading. Various parameters such as sliding speeds, track diameter, running time, and loads applied were set on the software that was linked to the tester machine. Then, the experiment was carried out. For the first set of experiments, nine sets of test run were conducted with three different values of normal load. For each load applied the test run was repeated three times as shown in Table 2.1. The running time was fixed for every test run at 60 min each. The test was conducted under dry surface conditions at room temperature ranging from 23 to 27 °C with a relative humidity (RH) of 70 (±5)%.

For the second set of experiments, six sets of test run were conducted with three values of normal load and two values of sliding speed as shown in Table 2.2. The running time was fixed for every test run at 60 min each and the same

**Table 2.1** Parameter for first test run.

| Specimen | 1 | 2 | 3 | 4 | 5 | 6 | 7 | 8 | 9 |
|---|---|---|---|---|---|---|---|---|---|
| Load | 50 | 50 | 50 | 100 | 100 | 100 | 150 | 150 | 150 |
| Speed | 500 | 500 | 500 | 500 | 500 | 500 | 500 | 500 | 500 |

**Table 2.2** Parameter for second test run.

| Specimen | 1 | 2 | 3 | 4 | 5 | 6 |
|---|---|---|---|---|---|---|
| Load | 50 | 100 | 150 | 50 | 100 | 150 |
| Speed | 100 | 100 | 100 | 300 | 300 | 300 |

condition as in set 1, dry surface conditions at room temperature ranging from 23 to 27 °C with RH of 70 (±5)%, were maintained.

Tensile test was performed to determine the maximum tensile load that a bone can withstand before it starts to crack. This test was carried out by using Instron Universal Test Machine (Model No 3367) at the Applied Mechanic Laboratory, School of Mechanical Engineering, USM, Malaysia.

Firstly, the specimen was prepared using a turning machine according to ASTM C565. The dimensions were derived from ASTM standard. The ratio $d/D$ should be around half and the parallel length of the narrow section should be at least three times the size of the gauge diameter d. The radius of curvature should be large enough to prevent the stress concentration on that section. The length $D$ should be one quarter of the overall length of the specimen as shown in Figure 2.6. For the fabrication of the specimen, two different cutting tools are needed: a facing tool for the radius $d$ and $D$ and a round nose for the curvature radius. Secondly, the fixture was installed with the speed of the moving platform set to 0.025 mm $\text{min}^{-1}$. The program was set to increase the load applied in the scale of 0.01 N.

Next, a nano-indenter was used to characterize the low impact fracture of the materials. The nano-indenter can evaluate <100 nm across and <5 mm in thickness. Test methods include indentation for comparative and quantitative hardness determination and scratching for evaluation of wear resistance and thin film adhesion. This test was carried out by using the Nano-indenter Test Machine that is available in the Microscopic and Micro Analysis Lab, School of Mechanical Engineering, USM, Malaysia, as shown in Figure 2.7. The sample was indented at

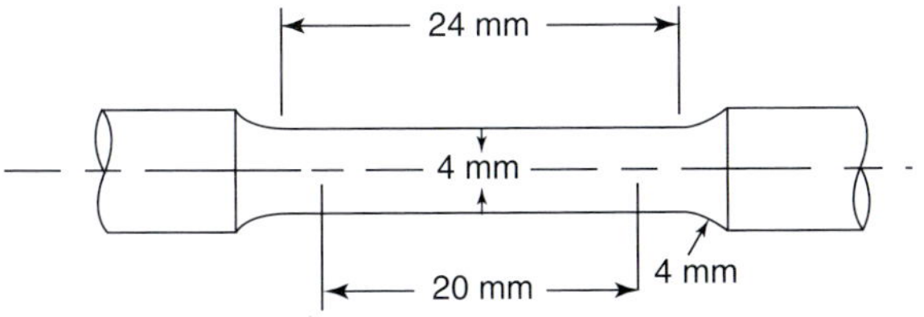


**Figure 2.6** The dimensions of the specimen as specified in ASTM C565.

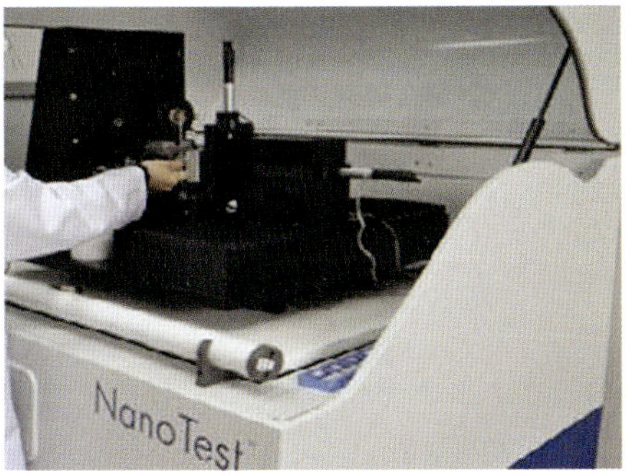


**Figure 2.7** Nano-indenter test machine and setup.

five points with an initial load of 0.03 mN and loading rate of 0.4 nM $s^{-1}$. Then, the values were averaged to get the optimum value.

SEM is a scanner focusing an electron beam for imaging purposes to obtain the surface topography. An embedded program known as energy dispersive X-ray analysis (EDX) is an X-ray technique that enables simultaneous imaging of the identified area of interest. Elemental mapping of a sample and image analysis are made possible with this technique.

## 2.3 Results and Discussion

Three different normal loads were applied at three different sliding speeds to different ranges of a normal healthy cow's bone to simulate the simplified Taguchi's experimental methods with low, medium, and high points. Table 2.3 shows the weight loss and average friction coefficient for various sliding speeds and load measured using a PODT.

From Figure 2.8, it is seen that the highest wear rate is produced by a sample of 500 rpm with a load of 150 N, and the lowest is at 100 rpm with a load of 50 N. Theoretically, the wear rate increases as the sliding speed and load increase. This was supported by Davim and Marques [21], who concluded that the higher sliding speed and contact stress exerted a great effect on the wear of the bone. The plotted results are slightly different as shown by the data plotted for 300 rpm with a load of 150 N followed by 500 rpm with a load of 50 N. This is because by increasing the normal applied load, more pressure was exerted on the sample with a sliding speed of 500 rpm whereby lesser force was exerted at 100 rpm.

The experiment was carried out under dry surface conditions at room temperature ranging from 23 to 27 °C with RH of 70(±5)%. The sliding speeds used in this experiment were 100, 300, and 500 rpm. Figures 2.9 and 2.10 show the variation of weight loss and COF with speed.

Figure 2.9 shows the effect of variation of sliding speed on weight loss on the bone. It can be observed that the weight loss increased as the speed increased.

**Table 2.3** General results of wear and COF tests on abrasion tribometer rig.

| Load (N) | Speed (rpm) | Weight loss (g) | Friction coefficient (COF) |
|---|---|---|---|
| 50 | 100 | 0.0163 | 0.2862 |
| 50 | 300 | 0.0301 | 0.3153 |
| 50 | 500 | 0.12939 | 0.28065 |
| 100 | 100 | 0.024 | 0.3051 |
| 100 | 300 | 0.0468 | 0.2681 |
| 100 | 500 | 0.1223 | 0.2725 |
| 150 | 100 | 0.0555 | 0.3033 |
| 150 | 300 | 0.0506 | 0.2522 |
| 150 | 500 | 0.18587 | 0.2120 |

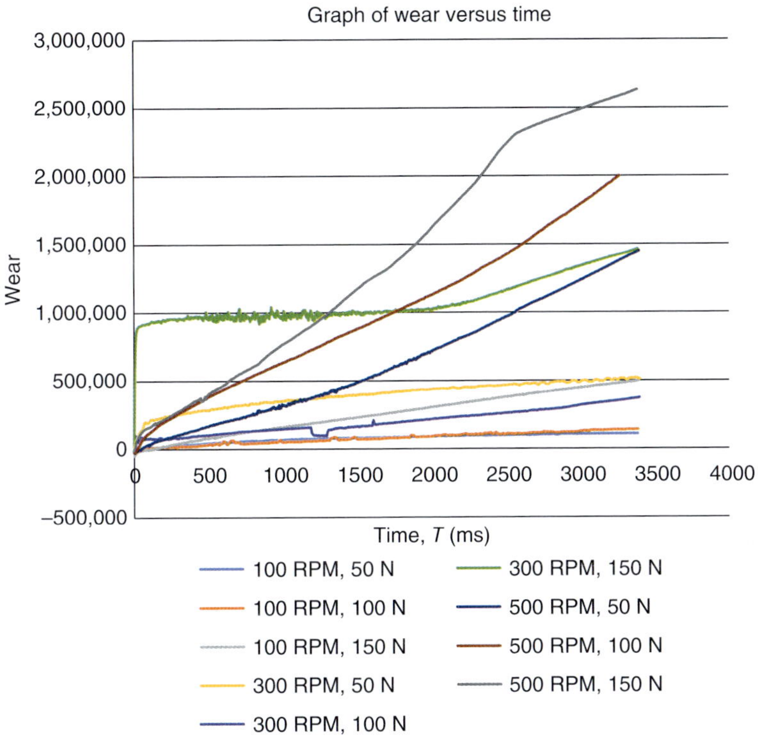


**Figure 2.8** Variation of wear rate in time series.

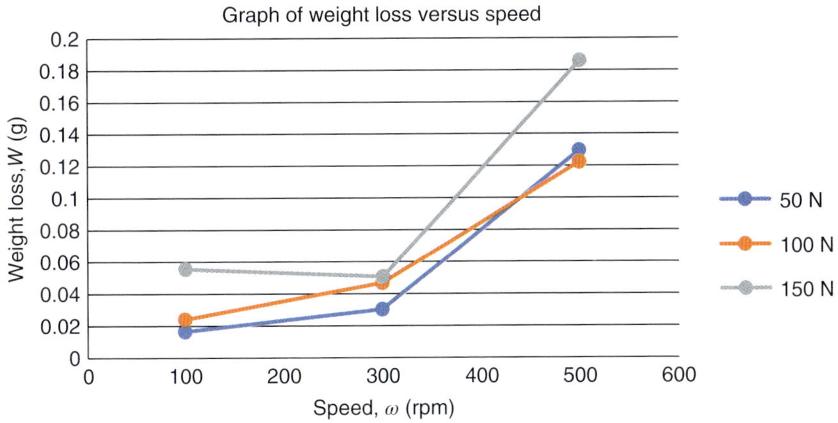


**Figure 2.9** Variation of weight loss with speed.

Previous literature reported that with higher sliding speeds, the rate of strain also increased. Hence, the brittleness of the worn material also increased and weight loss increased. In Figure 2.10, the variation of COF decreased as the speed increased. This can be clearly seen from the line for 150 N. The highest average COF resulted from a low sliding speed, which is 0.3533. This finding is validated

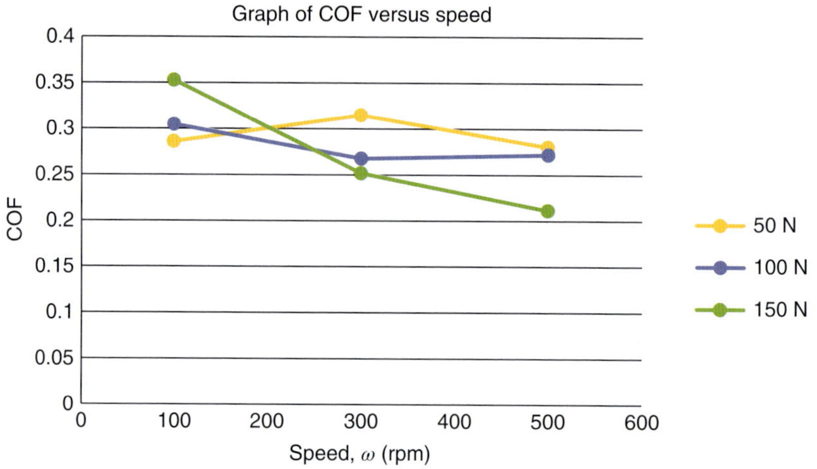


**Figure 2.10** Variation of coefficient of friction (COF) with speed.

by Chowdhury et al. [22], who found that COF decreased with increasing sliding speed. It may be due to the change in the shear rate, which can affect the mechanical properties of the mating materials.

Similarly, the experiment was carried out with varying loads under dry surface conditions at room temperature ranging from 23 to 27 °C with RH 70 (±5)%. The loads used in this experiment were 50, 100, 150 N. Figures 2.11 and 2.12 show the variation of weight loss and COF against load.

Figure 2.11 shows the variation of weight loss against applied load. When the load increased, the contact stress also increased, and thus the weight loss increased. Figure 2.12 shows COF as a function of the normal load at 100, 300, and 500 rpm of sliding speed. COF is inversely proportional to the normal load. Thus, the friction force and friction coefficient decreased, as agreed by Zhu et al. [12].

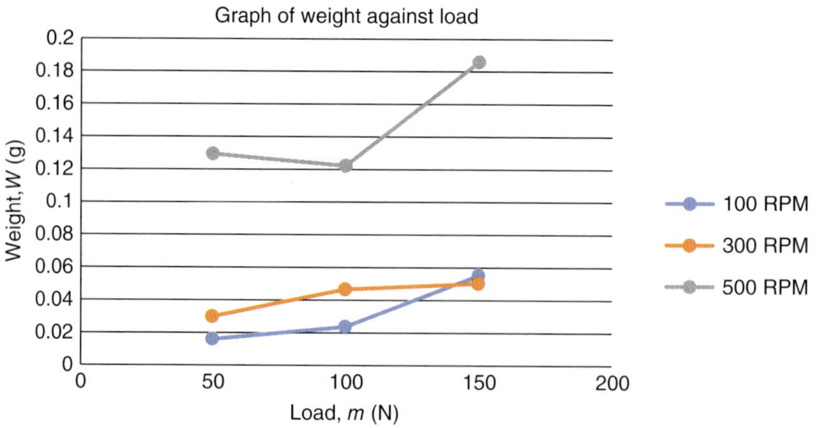


**Figure 2.11** Variation of weight loss with load.

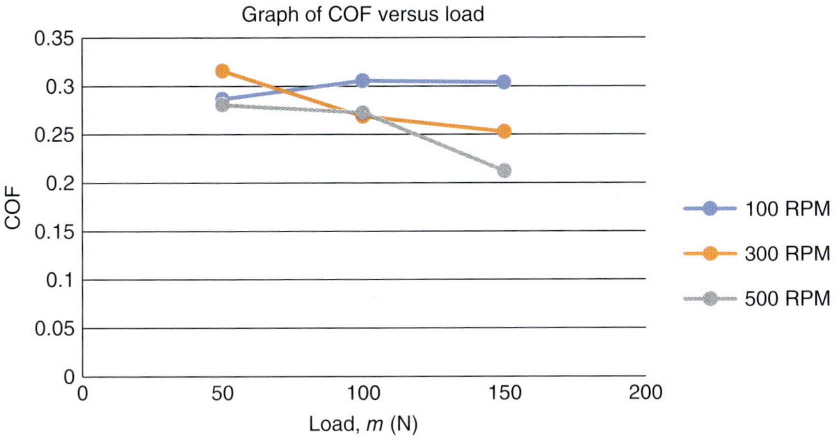


**Figure 2.12** Variation of COF with load.

Re-plotting the data from Figures 2.9–2.12 in Figure 2.13, it can be seen that COF and weight loss are interconnected to each other. In general, weight loss increased as COF decreased. The range of weight loss was between 0.0163 and 0.186 g, whereas the range of COF was between 0.21 and 0.31. The highest weight loss from the sample at 500 rpm with load of 150 N also gave the lowest value of COF. The lowest weight loss was from the sample at 100 rpm with load of 50 N but the highest COF resulted from the sample with 500 rpm with load of 50 N. This is as agreed by Asuke et al. [23], who stated that when load is low, the wear loss is quite small and increases with increase in applied load. It is quite natural for the wear rate to increase with applied load. A similar trend was observed independently for different wear distances as a function of load and speed.

The SEM optical microscopic images of worn surfaces of bone are presented in Figure 2.14a–i. Scratch marks are clearly seen in all images for the tests carried

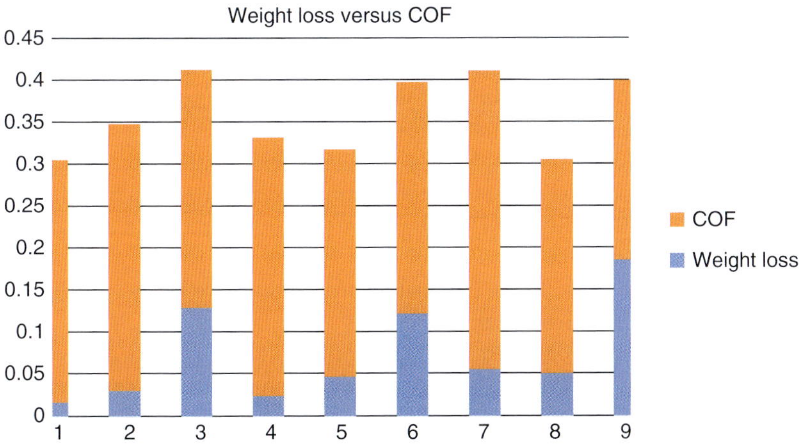


**Figure 2.13** Interrelation between weight loss and COF.

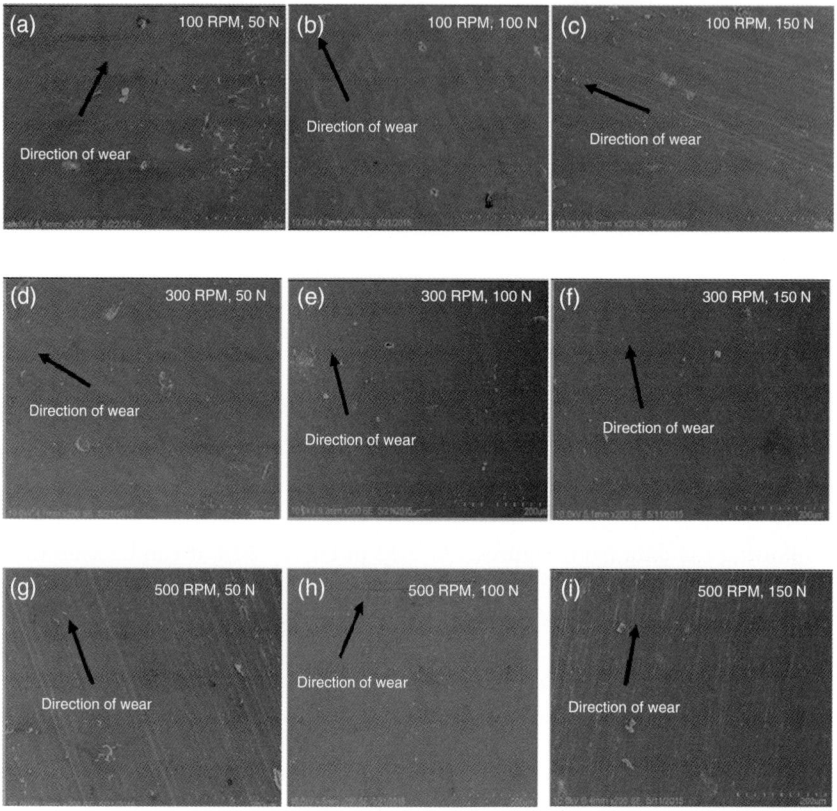


**Figure 2.14** (a–i) The image of the bone under 200× magnification.

out, which shows that the surfaces have experienced abrasive wear. By comparing the entire set of images, the texture of the groove is deeper and coarser as the normal load and sliding speed increase.

Figure 2.15 shows that the fracture starts to propagate when higher load shear stress acting parallel to the surface of the bone is exerted on the bone. With higher load being exerted on the sample, the fracture starts to spread all over the surface, and eventually the samples start breaking.

Based on the EDAX analysis in Table 2.4, the amount of oxygen present in the sample is 37.91% of weight, which is the highest among all elements. The calcium present in the sample is about 25.97%. There are also other impurities found in the specimen such as carbon (C), sodium (Na), magnesium (Mg), phosphorus (P), and others (Figure 2.16). These impurities are present due to the contact of the specimen with the stainless steel pin.

Tensile test produced a load versus elongation curve, which was then converted into a stress versus strain curve. Since both the engineering stress and the engineering strain are obtained by dividing the load and extension by a constant, the load–extension curve will have the same shape as the engineering stress–strain curve. The stress–strain curve relates the applied stress to the resulting

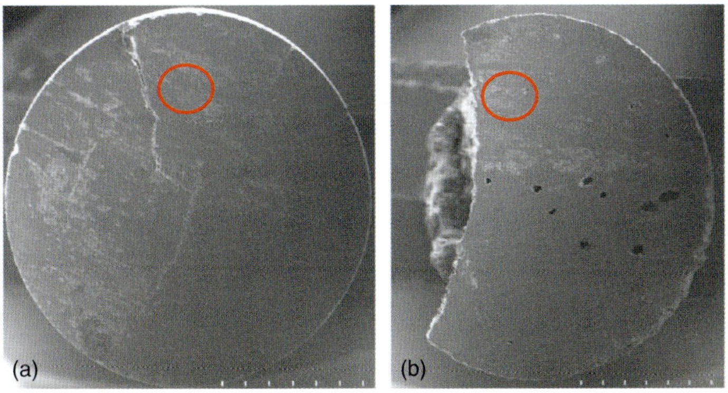

Figure 2.15 The image of bone at (a) 300 rpm and 100 N (b) at 500 rpm and 150 N.

Table 2.4 Chemical composition and properties of the bone.

| Element | Weight % | Atomic % | Net Int | Net Int Error |
|---|---|---|---|---|
| C K | 16.17 | 27.23 | 101.27 | 0.03 |
| O K | 37.91 | 47.91 | 348.96 | 0.01 |
| Na K | 0.76 | 0.67 | 31.88 | 0.07 |
| Mg K | 0.72 | 0.6 | 38.11 | 0.06 |
| P K | 14.96 | 9.76 | 729.58 | 0.01 |
| Nb L | 2.17 | 0.47 | 48.58 | 0.07 |
| Pd L | 1.33 | 0.25 | 22.04 | 0.21 |
| Ca K | 25.97 | 13.1 | 850.89 | 0.01 |

strain and each material has its own unique stress–strain curve. Figure 2.17 shows the graph of stress versus strain. The areas under this graph indicate the strength of the sample tested. Generally, strength shows the amount of energy that can be absorbed before the sample ruptures. The slope of the stress–strain curve shows the modulus of elasticity (Young's modulus). The modulus of elasticity is a measure of the stiffness of the material, but it only applies in the linear region of the curve. The value of Young's modulus obtained from this test was 11.167 GPa. The maximum stress and strain that the sample can withstand were 95.24 MPa and 0.033 75 mm/mm. Besides, Figure 2.18 shows the graph of load versus extension. These graphs present the elasticity of the tested sample. The sample can withstand a maximum load of 1196.86 N and the maximum elongation produced was 0.810 05 mm.

Hardness test was carried out in two conditions – one with coating with resin and another without the resin coating. The purpose of using resin was to fill up the cavity causing irregularity on the surface of the bone. This will ease the probe of nano-indenter machine to get a reading.

Bones without coating of resin were tested under dry room temperature with 8.0 N of maximum indenter load. There were five points of indention on the

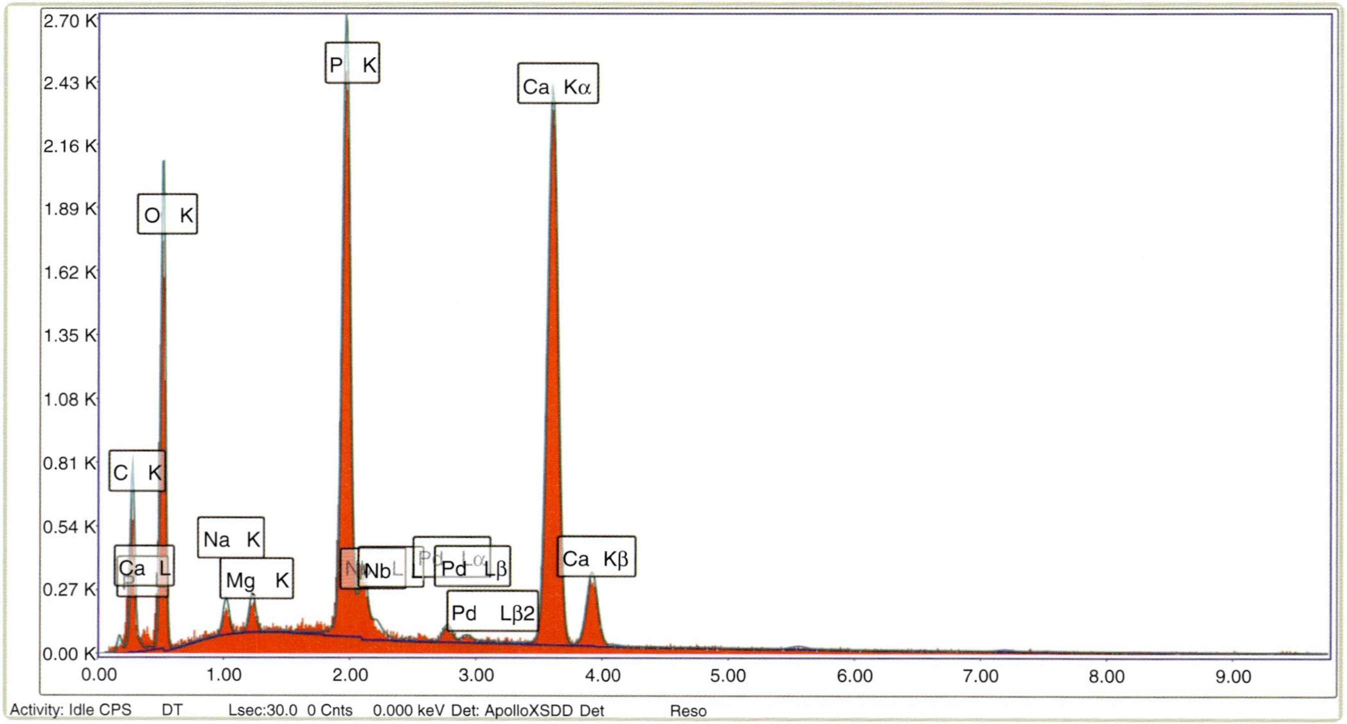


**Figure 2.16** The composition of bone.

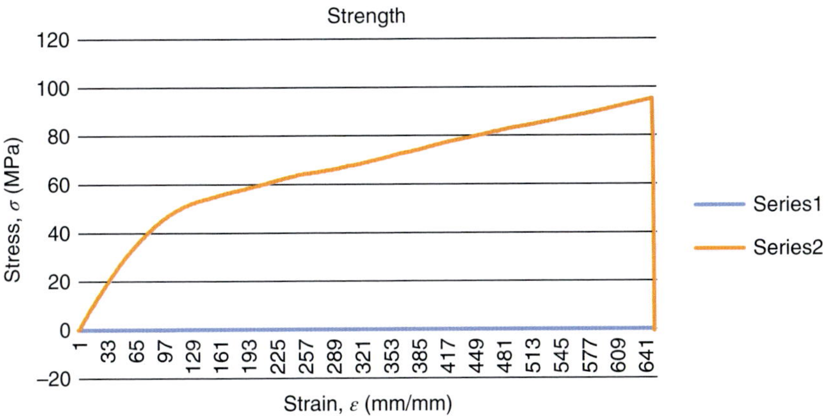


**Figure 2.17** The stress versus strain graph.

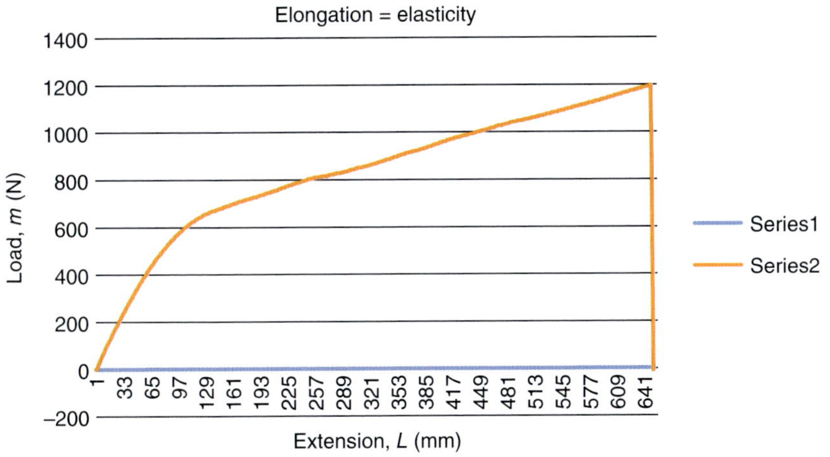


**Figure 2.18** Load versus extension graph.

sample. Each of the points was taken randomly, with about 200 nm distance between two points (Figure 2.19).

Bones with coating of resin were also tested under dry room temperature with 8.0 N of maximum indenter load. There were five points of indention on the sample as shown in Table 2.5. Each of the points was taken randomly at about 200 nm distance between two points (Figure 2.20).

The indention depth and hardness of the five points are tabulated in Table 2.6 with average 5449.555 nm and 0.0368 GPa.

The test was carried out without using a resin to coat the surface of bone. The indention depth and hardness of five points are tabulated in Table 2.5 with average 518.6915 nm and 1.647 068 GPa. From these two graphs, it can be concluded

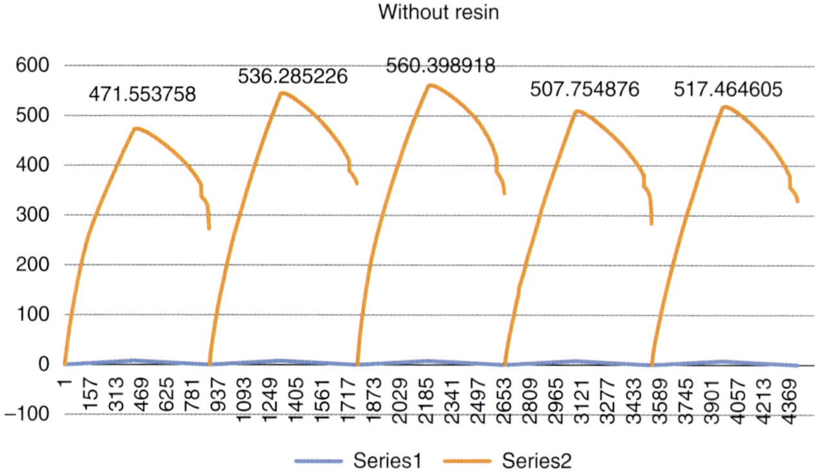


**Figure 2.19** Graph of load against indention depth (without resin coating).

**Table 2.5** Average indention depth (without resin).

| Peak | 1 | 2 | 3 | 4 | 5 | Average |
|---|---|---|---|---|---|---|
| Value | 471.5538 | 536.2852 | 560.3989 | 507.7549 | 517.4646 | 518.6915 |
| Hardness | 1.940889 | 1.498884 | 1.373054 | 1.726510 | 1.696002 | 1.647068 |

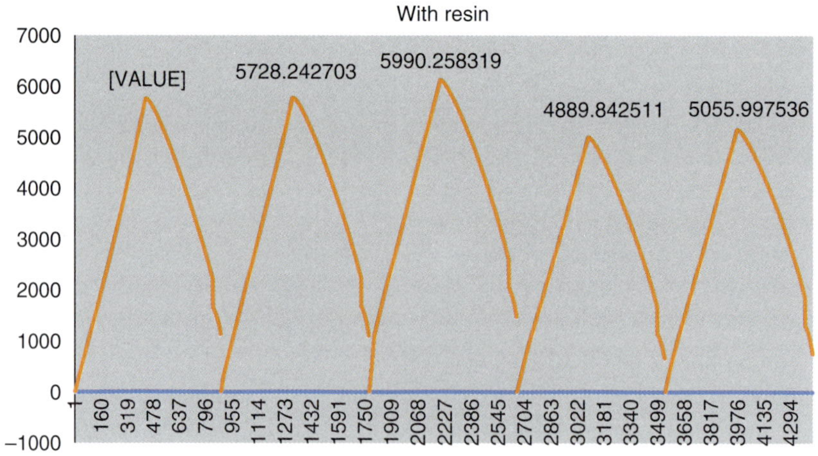


**Figure 2.20** Graph of load against indention depth (with resin coating).

that the hardness of the bones without using resin is better than that of the bone with resin coating. Theoretically, the hardness of the sample with coating is greater than that without coating. The function of the resin is to strengthen the bones, which is at variance with the results of this project.

**Table 2.6** Average indention depth (with resin).

| Peak | 1 | 2 | 3 | 4 | 5 | Average |
|---|---|---|---|---|---|---|
| Max depth | 5583.435 | 5728.243 | 5990.258 | 4889.843 | 5055.998 | 5449.555 |
| Hardness | 0.033704 | 0.033222 | 0.025964 | 0.048549 | 0.042553 | 0.0368 |

## 2.4 Conclusion

It can be concluded that weight loss and wear rate increase as the sliding speed and normal load increase. But, COF decreases when the sliding speeds and normal load increase.

1) From the SEM analysis it is seen that the worn surfaces become coarser with deeper grooves as the applied load and sliding speed increase. This is due to the increasing amount of material removed during the wear test.
2) EDAX analysis shows the chemical element composition of the bone. Results show that oxygen and calcium are the elements present in high amounts. There are also impurities present during the analysis.
3) From the tensile test, the value of tensile strength of the bone is 11.67 GPa.
4) Hardness test was carried out under two conditions for the sample: with resin coating and without resin coating. The value of hardness for the sample with resin coating is 0.3 GPa and for that without coating is 1.19 GPa. The difference between these values is due to the presence of resin.

## References

1 Encyclopaedia Britannica (2006). Human bone.
2 Dumitrescu, A.L. (2011). Bone grafts and bone graft substitutes in periodontal therapy. In: *Chemicals in Surgical Periodontal Therapy* (ed. A.L. Dumitrescu). Berlin: Springer-Verlag.
3 Adam, C.J. and Swain, M.V. (2011). The effect of friction on indenter force and pile-up in numerical simulations of bone nanoindentation. *Journal of the Mechanical Behavior of Biomedical Materials* 4 (7): 1554–1558. doi: 10.1016/j.jmbbm.2011.03.026.
4 Rodan, G.A. (1992). Introduction to bone biology. *Bone* 13 (Suppl. 1): S3–S6. doi: 10.1016/S8756-3282(09)80003-3.
5 Qin, Q.H. (2013). *Mechanics of Cellular Bone Remodeling: Coupled Thermal, Electrical, and Mechanical Field Effects*. CRC Press.
6 Vanasse, A., Dagenais, P., Niyonsenga, T. et al. (2005). Bone mineral density measurement and osteoporosis treatment after a fragility fracture in older adults: regional variation and determinants of use in Quebec. *BMC Musculoskeletal Disorders* 6: 33–33. doi: 10.1186/1471-2474-6-33.
7 Jones, M.H. and Scott, D. (ed.) (1983). *Industrial Tribology: The Practical Aspects of Friction, Lubrication, and Wear*. New York: Elsevier Scientific Publishing Company.
8 Khruschov, M.M. (1974). Principles of abrasive wear. *Wear* 28 (1): 69–88. doi: 10.1016/0043-1648(74)90102-1.

9 McKellop, H., Clarke, I., Markolf, K., and Amstutz, H. (1981). Friction and wear properties of polymer, metal, and ceramic prosthetic joint materials evaluated on a multichannel screening device. *Journal of Biomedical Materials Research* 15: 619.
10 Yu, H.Y., Cai, Z.B., Zhou, Z.R., and Zhu, M.H. (2005). Fretting behavior of cortical bone against titanium and its alloy. *Wear* 259: 910–918.
11 Q.-L., W. and Ge, S.-R. (2007). Comparison of biotribology of swine compact bone against UHMWPE. *Journal of China University of Mining Technology* 17 (1): 133–137.
12 Zhu, B., He, X., and Zhao, T. (2015). Friction and wear characteristics of natural bovine bone lubricated with water. *Wear* 322–323: 91–100. doi: 10.1016/j.wear.2014.10.013.
13 Ghosh, S., Choudhury, D., Das, N.S., and Murphy, B.P. (2014). Tribological role of synovial fluid compositions on artificial joints – a systematic review of the last 10 years. *Lubrication Science* 26 (6): 387–410.
14 Lizhang, L., Fisher, J., Jin, Z. et al. (2010). The effect of contact stress on cartilage friction, deformation and wear. *Proceedings of the Institution of Mechanical Engineers, Part H: Journal of Engineering in Medicine* 225 (5): 461–475.
15 T, M., Higaki, H., Sawae, Y. et al. (1998). Adaptive multimode lubrication in natural synovial joints and artificial joints. *Proceedings of the Institution of Mechanical Engineers, Part H: Journal of Engineering in Medicine* 212 (1): 23–35.
16 Bowman, S.M., Zeind, J., Gibson, L.J. et al. (1996). The tensile behavior of demineralized bovine cortical bone. *Journal of Biomechanics* 29 (11): 1497–1501. doi: 10.1016/0021-9290(96)84546-5.
17 Zhang, W., Tekalur, S.A., Baumann, M., and McCabe, L.R. (2013). The effects of damage accumulation on the tensile strength and toughness of compact bovine bone. *Journal of Biomechanics* 46 (5): 964–972. doi: 10.1016/j.jbiomech.2012.12.005.
18 An, Y.H. and Draughn, R.A. (1999). *Mechanical Testing of Bone and the Bone-Implant Interface*. Taylor & Francis.
19 Zysset, P.K., Edward Guo, X., Edward Hoffler, C. et al. (1999). Elastic modulus and hardness of cortical and trabecular bone lamellae measured by nanoindentation in the human femur. *Journal of Biomechanics* 32 (10): 1005–1012. doi: 10.1016/S0021-9290(99)00111-6.
20 Dall'Ara, E., Schmidt, R., and Zysset, P. (2012). Microindentation can discriminate between damaged and intact human bone tissue. *Bone* 50 (4): 925–929. doi: 10.1016/j.bone.2012.01.002.
21 Davim, J.P. and Marques, N. (2004). Dynamical experimental study of friction and wear behaviour of bovine cancellous bone sliding against a metallic counterface in a water lubricated environment. *Journal of Materials Processing Technology* 152 (3): 389–394. doi: 10.1016/j.jmatprotec.2004.04.420.
22 Chowdhury, A.M., Nuruzzaman, D., Mia, A., and Rahaman, M. (2012). Friction coefficient of different material pairs under different normal loads and sliding velocities. *Tribology in Industry* 34 (1): 18–23.
23 Asuke, F., Abdulwahab, M., Aigbodion, V.S. et al. (2014). Effect of load on the wear behaviour of polypropylene/carbonized bone ash particulate composite. *Egyptian Journal of Basic and Applied Sciences* 1 (1): 67–70. doi: 10.1016/j.ejbas.2014.02.002.

# 3

# Tribological Behaviors of Glass Fiber with Fillers Reinforced Hybrid Polymer Composites

*T. P. Sathishkumar*

*Kongu Engineering College, Department of Mechanical Engineering, Erode 638060, Tamil Nadu, India*

## 3.1 Introduction

Synthetic fibers and fillers are extensively used as good reinforcement materials in various polymeric composites. Prodigious reinforcing materials along with synthetic fibers are abundant sources of natural fillers to produce hybrid fibers reinforced polymeric composites. The synthetic fibers are classified into glass, kevlar, and carbon fibers. They have high strength, stiffness, length-to-weight ratio, and good fire resistance. These fibers are high in cost and cause pollution when they are brought out. The various fillers are graphite, silicon graphite, polyetheretherketone, polymers, and silicon carbide, which when gently mixed with fibers reinforced composite help increase the mechanical properties. Over the last two decades the usage of fillers in glass fiber polymeric composites has widely increased and this reduces the weight and cost of industrial products. In fact, they have many advantages such as light weight, low cost, high specific strength, better insulation, and flexibility [1]. The study of mechanical and tribological properties of fillers and glass fiber reinforced polymer composites is essential for optimizing filler parameters such as filler diameter, filler volume and weight fractions, and matrix content. The concepts of "green tribology" are introduced by the World Tribology Congress (WTC). The aim is to save energy and materials by introducing new equipment in the market that reduces the need for fossil fuels. In order to meet the WTC concepts, fillers are introduced in glass fiber polymer composites [2]. Tribological study on the new materials is essential to satisfy the needs of real-time applications.

Machine and structural failures occur frequently because of the sliding of components due to adhesive or abrasive contact, which results in loss of more material. This situation is so critical that it needs investigation on the friction and wear behaviors of fillers in glass reinforced polymer materials. This investigation is normally carried out on the wear test rig in the laboratory before the material is used in applications. Tribo properties such as rate of wear, specific wear rate, and coefficient of friction (COF) are investigated by varying test parameters including counterpart materials, velocity and distance, applied normal loads,

*Synthesis and Tribological Applications of Hybrid Materials*, First Edition.
Edited by Mohammad Jawaid, Rajini Nagarajan, Jacob Sukumaran, and Patrick De Baets.

testing environments, and testing orientations with respect to orientation of reinforcements. Glass fillers reinforced hybrid composites are prepared by various preparation methodologies [1] such as the injection molding process, compression molding process, and vacuum molding process.

## 3.2 Wear and Mechanisms of Wear

Wear is the loss of material that takes place because of different reasons. The mechanism of wear differs for different materials. In order to enhance the life of a material, the wear mechanism should be studied with different forms of wear. They are adhesion, abrasion, delamination, surface fatigue, and tribochemical reaction wear. Usually, physical and chemical wear takes place by contact of two bodies. Major wear and friction behaviors are examined by the following two wears.

### 3.2.1 Adhesion Wear

The two surfaces mate to generate high pressure, which leads to the formation of a cold weld at junctions due to the plastic deformation of bodies. Then the surfaces begin to move against each other, which leads to material loss at the junction if the material has higher shear strength leading to less wear.

### 3.2.2 Abrasive Wear

Abrasive wear is common for all forms of wear and constitutes 63–73% of wear for all industrial components. The surface of the material wears against the hard counterface, which leads to material loss based on the nature of the interface. There are two types of abrasive wears such as two body and three body abrasion. The two hard surfaces slide to form two body abrasive wear (Figure 3.1a). For example, the automobile break pad dry-slides on the break drum for braking and the two gears mate together. This wear occurs through single pass and multi-pass of the material surfaces. Normally, the wear rate is lower in multi-pass than in

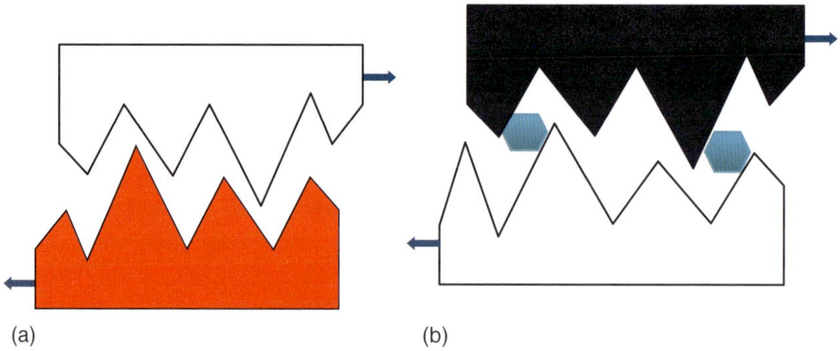


**Figure 3.1** (a) Two body and (b) three body abrasive wear.

single pass. When the hard particles slide as well as roll in between, the two hard surfaces form three body abrasive wear (Figure 3.1b), and the rate of wear in this mechanism is lower than that in two body wear. Therefore, 90% of the wear occurs by rolling contact than by sliding.

## 3.3 Tribo Wear Test Methods

Wear and friction tests are conducted using various testing methods according to international standards and industrial needs.

### 3.3.1 Wear and Friction Test Using Pin-on-disk

The pin-on-disk (POD) or block-on-disk (BOD) tribo testing machine is shown in Figure 3.2 [3]. The test specimen is fitted into a vertical specimen holder. A dead weight is added on the specimen holder, which is the normal applied load of the wear test. The specimen holder with specimen is kept on a rotating counterface, which is made up of stainless steel and is rotated by an electric motor. A long beam has the specimen holder at the left end and weight balance at the right end; it is pivoted on a machine structure between the two ends. The wear test was carried out according to ASTM G99-05 standard. The hybrid composite specimen of 10 mm × 10 mm × 20 mm was fitted into the specimen holder and placed on the rotating counterface. The counterface velocity and sliding distance were varied according to the material and the environmental conditions. The wear and friction behaviors were measured from the experiments.

The COF ($\mu$) was calculated using Equation 3.1. It is the ratio of the measured frictional force to the normal applied load. The measured friction force and applied normal load are expressed in newton.

$$\text{Coefficient of friction}(\mu) = \frac{\text{Measured frictional force}}{\text{Normal applied load}} \tag{3.1}$$

The weight loss for each specimen was measured before and after the wear test using a 0.1 mg digital weight indicator. The specific wear rate ($W_s$) was calculated using the following relation (Equation 3.2).

$$\text{Specific wear rate}\left(W_s\right) = \frac{W_I - W_F}{\rho \times F_N \times D} \tag{3.2}$$

where

$W_s$ = specific wear rate ($mm^3\,N^{-1}\,m^{-1}$)
$W_I$ = initial weight of the specimen (kg)
$W_F$ = final weight of the specimen (kg)
$\rho$ = density of the composite ($kg\,mm^{-3}$)
$F_N$ = normal applied load (N)
$D$ = sliding distance of specimen (m)

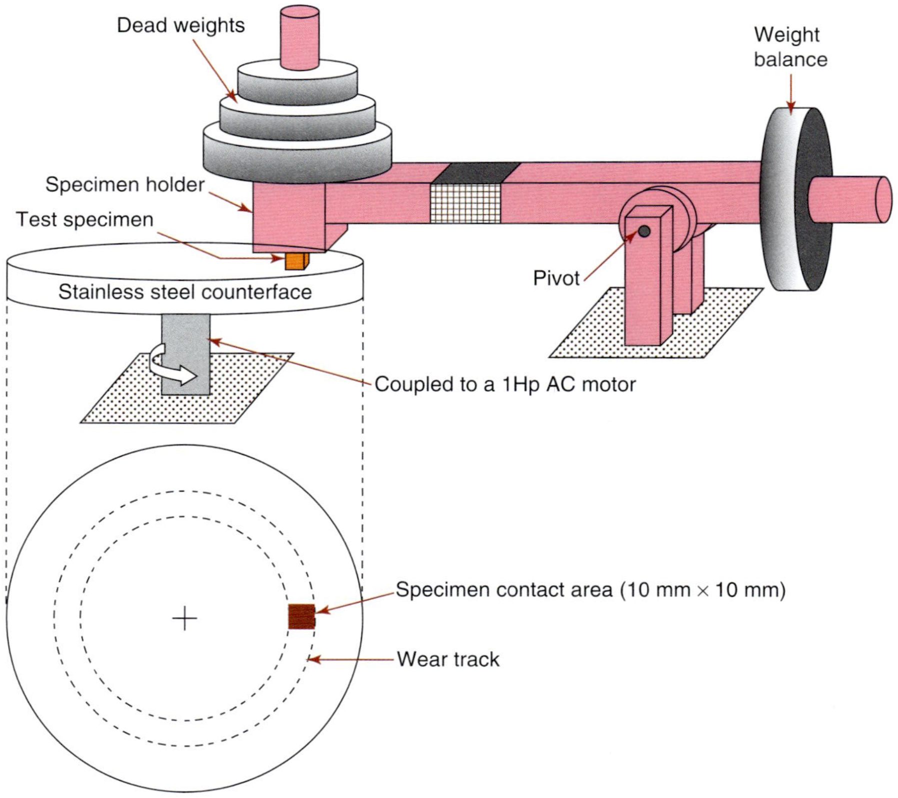


**Figure 3.2** Schematic of the pin-on-disk tribo test machine.

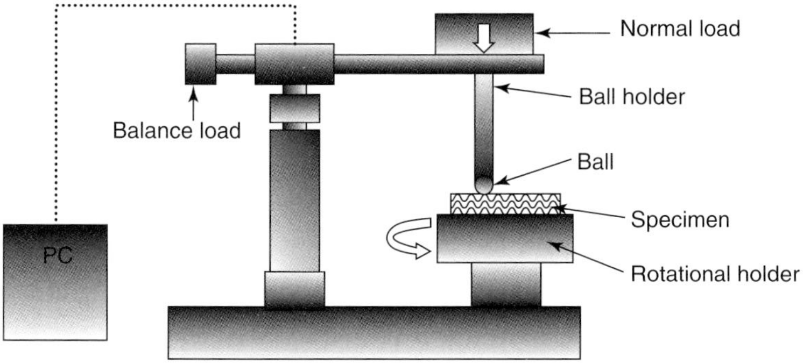


**Figure 3.3** Schematic of the ball-on-disk tribo testing machine. *Source:* Thongsang et al. 2012 [4]. Reproduced with permission from Elsevier.

### 3.3.2 Wear and Friction Test Using Ball-on-disk

The schematic of a BOD tribo testing machine is shown in Figure 3.3 [4]. The metal ball is made of steel, stainless steel, chromium, aluminum alloy, etc. and is fixed at the bottom side of the ball holder, which acts as the counterface (i.e. counterpart). The beam consists of the ball holder at the right end where the normal load is applied on it for tribo test and the balance load at the left end. The beam is pivoted on the machine structure column. The wear specimen is kept on the rotating holder, which is controlled by an electrical motor. During testing, the static metal ball is placed on the specimen. By varying testing parameters such as sliding velocity of the specimen, sliding distance, and normal load, the specimen is rotated on the steel ball surface and then the frictional force is measured. The tribo testing machine has a multi-channel electronic unit and is connected to a personal computer (PC) for data acquisition.

## 3.4 Tribo Characterization Hybrid Polymer Composites

### 3.4.1 Polyamide 6 and HDPE Glass Fiber Reinforced Hybrid Composites

Polyamide 6 (PA6) and high density polyethylene (HDPE) are blended with the help of maleic anhydride modified polypropylene (MAgPP) as a compatibilizer [5]. The MAgPP added is 5 wt%. The blender is prepared by mixing the PA6 and HDPE in the ratio of 80 : 20 wt%. Glass fibers of 5, 10, and 15 wt% are added to the polyblend to prepare the glass fiber hybrid composites. The glass fiber is 17 μm in diameter and 5 mm in length. Initially, the polymers and compatibilizer are dried at 85 °C for 2 days to avoid the effect of hydrolyzing and plasticization. The dried polymers are carefully mixed and put on the hopper in an extruder machine. They are extruded through a twin screw extruder barrel at various heating zones of 180, 235, 240, 245, and 240 °C respectively to prepare the cylindrical rods. The extruded rods are quenched in water and pelletized. The reciprocating screw injection molding process (IMP) is used to make composite samples from the polyblend pellets for tribology testing. Moreover, the glass fiber (GF) is added with polyblend pellets during IMP to prepare glass fiber reinforced hybrid composites samples. The barrel temperatures of the injection molding machine are 215, 220, and 240 °C at the three zones, and 230 °C at the nozzle. The mold temperature is 90 °C.

Composite pins of 6 mm diameter and 16 mm length are taken for friction and wear test using the POD machine. Before the test, the wear surface of the pins is abraded using a 600 SiC grade sheet. This will ensure maximum contact between the pin and the counterface. The counterface disk material is AISI 02 tool steel with 75 mm diameter and 5 mm thickness. This may be abraded using 350 SiC grade sheet to maintain a surface roughness of 0.10–0.11 μm. The counterface is rotated by an electrical motor with sliding velocity of $1\,m\,s^{-1}$ and applied load of 19.6 N. The wear track diameter is 65 mm.

Figure 3.4 shows a plot of the COF and wear loss versus sliding distance of 80 wt% PA–20 wt% HDPE polyblend with and without GF in hybrid composites. The COF have some variations in the beginning stage of sliding. The pure polymer shows reduction in COF with sliding distance. The addition of 5% GF to the composite shows higher COF when compared to 10% and 15% GF. The composites with 10% and 15% GF have steady state wear up to 5 km sliding distance. However, there are no significant changes in 0% GF reinforcement. But the COF and wear loss are low for 10% GF hybrid composites. Figure 3.5 shows the worn surface morphology. Figure 3.5a shows that the transfer film forms on the steel counterface, and this formation is uniform and thin. This may avoid aggressive damage of the pin surface and reduce the wear rate. Figure 3.5b shows the wear track of the polyblend with 15% GF reinforced hybrid composites on the counterface. The transfer film is formed with shiny spots. It shows the presence of

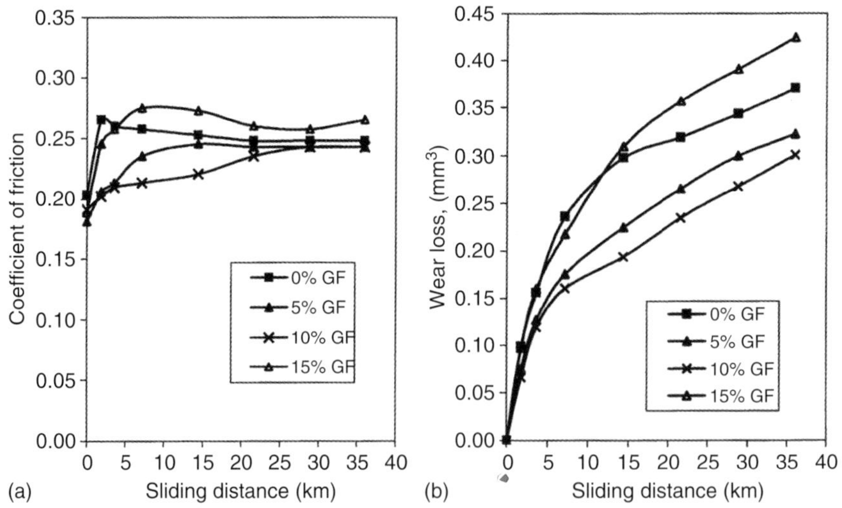


**Figure 3.4** (a) The coefficient of friction and (b) wear loss for 80 wt%/PA–20 wt% in HDPE polyblend with GF. *Source:* Palabiyik and Bahadur 2002 [5]. Reproduced with permission from Elsevier.

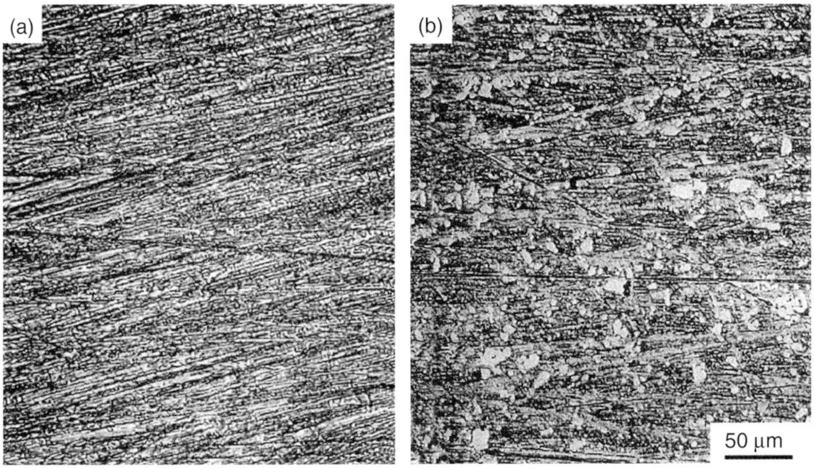


**Figure 3.5** Optical micrographs of the steady-state transfer films formed by 80 wt% PA6/20 wt% HDPE polyblend with the compositions (a) PA6 and HDPE and (b) 15% GF. *Source:* Palabiyik and Bahadur 2002 [5]. Reproduced with permission from Elsevier.

glass fibers in transfer materials. The presence of 15% glass fiber in the polyblend slightly increases the rate of wear. But polyblend with 5% and 10% GF shows lower wear rate as compared to control and that with 15% GF.

Solid lubricants such as polytetrafluroethylene (PTFE) and ultra-high molecular weight polyethylene (UHMWPE) are taken to prepare the glass fiber reinforced PA6 composites [6]. The weight content of glass fiber and PA6 is 15/85%. The glass fiber with different solid lubricants is mixed and extruded in twin screw extruders to prepare the pellets. The processing temperature in extruder barrels is 210, 220, 230, and 240 °C respectively at nine zones. The prepared pellets are injection molded to prepare the samples. BOD machine is used to perform the tribological behavior tests (Figure 3.3). The diameter of the chromium ball is 9.5 mm and it act as the counterface. The sample dimensions for wear test are $25 \times 25 \times 15\,\text{mm}^3$ and it is fixed in tribo machine fixture. Before the test, the specimen surfaces are cleaned ultrasonically with acetone. The tribo test is conducted with two loads (40 and 100 N) and velocities (200 and 1500 rpm). The COF is calculated using Equation 3.1 and the wear rate ($\text{mm}^3\,\text{N}^{-1}\,\text{m}^{-1}$) is calculated using Equation 3.3.

$$\text{Wear rate} = \frac{B}{Ld}\left[\frac{\pi r^2}{180}\sin\left(\frac{b}{2r}\right) - \frac{b}{2}\sqrt{r^5 - \frac{b^2}{4}}\right] \tag{3.3}$$

where

$B$ = the trace of friction (10 mm)
$r$ = the semi-diameter of the chromium steel ball (mm)
$b$ = the width of the wear trace (mm)
$L$ = the load (N)
$d$ = the sliding distance (m)

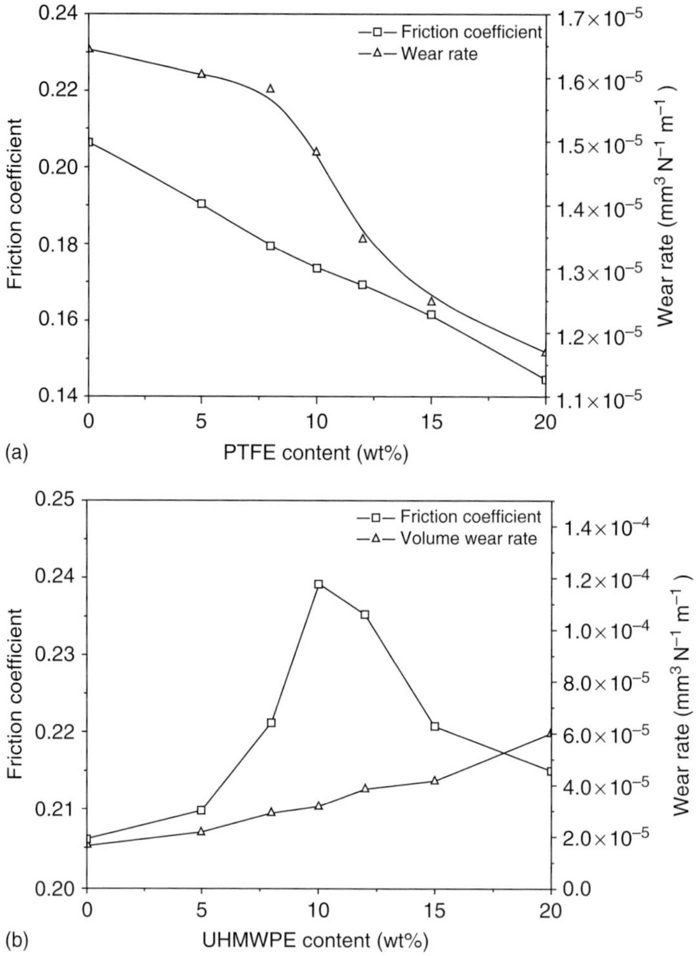


**Figure 3.6** Friction coefficient and wear rate of GF/PA6 composite by varying (a) PTFE and (b) UHMWPE (testing parameters: 40 N and 200 rpm). *Source:* Li et al. 2013 [6]. Reproduced with permission from Elsevier.

Figure 3.6 shows the differences in friction coefficients and wear rate for two different solid lubricants presented in the GF/PA6 composites at testing parameters of 40 N and 200 rpm. The pure PA6 composite has 0.2111 $\mu$ and 2.769 wear rate. The figure shows that the incorporation of glass fiber in the PA6 with solid lubricants can reduce the friction coefficients and increase the wear resistance. Incorporation of PTFE in the composites gradually decreases the friction coefficient and wear rate. The best tribo properties occur in GF/PA6 composites with 20% PTFE. By reducing the glass fiber fracture a low friction coefficient can be obtained. A third body transfer film is formed on the counterface by wearing of PTFE. During the sliding process, the banded crystalline PTFE can easily transfer due to low surface energy when compared to PA6 transfer. In addition, PTFE molecule chain breakage takes place in the chemical bonds of C—C or C—F. The breakage of C—F bond generates F-ions, which will react with the counterface to form a strong transfer film. Hence, the reduction in rate of wear occurs because of difficulty in breaking the formed transfer film. In Figure 3.6, the wear coefficient and wear rate are continuously increased by incorporating the UHMWPE in GF/PA6 composites. The UHMWPE solid lubricant does not show a positive effect on tribological properties because weaker interfacial adhesion occurs between the PA6 and UHMWPE, and it leads to more fracture of glass fibers. The high level of fiber debris disrupts the soft transfer film formation on the counterface. This may increase the wear rate and friction coefficient. Figure 3.7 presents the effects of friction and wear properties of both solid lubricants incorporated in GF/PA6 composites. The solid lubricants are loaded at 15 wt%. There are some beneficial impacts on the friction coefficients but not much improvement on wear rate. Overall, the GF/PA6 composite with 12 wt% PTFE/3 wt% UHMWPE shows the lowest friction coefficients, but the wear rate is slightly higher than that with 15 wt% PTFE alone. Increasing the PTFE proportion improves the wear and friction properties, and the combination of both solid lubricants exhibits better wear results than UHMWPE alone. Therefore, the removal of PA6 is reduced during tribo test with incorporation of PTFE. This occurs through some important mechanisms such as material transfer, viscous flow of polymer, and fatigue wear. The very high bonding that exists between PA6 and PTFE hampers materials transfer.

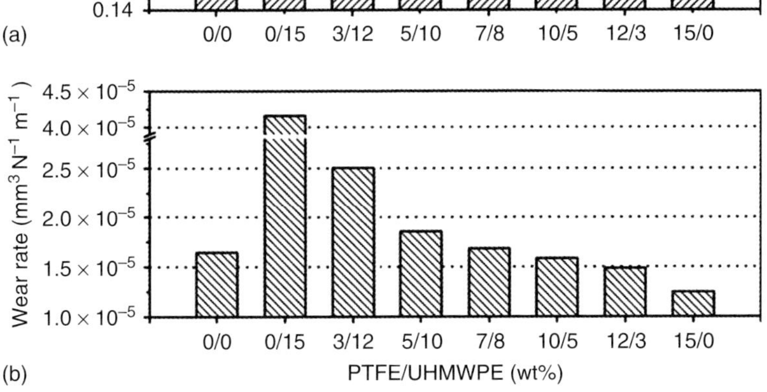


**Figure 3.7** (a) Friction coefficient and (b) wear rate of GF/PA6 composite with both PTFE and UHMWPE (testing parameters: 40 N and 200 rpm). *Source:* Li et al. 2013 [6]. Reproduced with permission from Elsevier.

### 3.4.2 Silicon Carbide, Graphite Particle, and Glass Fiber Reinforced Hybrid Composites

The 7-mil E-glass fiber fabric and fillers of SiCp (15 μm) and Gr (15 μm) powders are used to prepare the hybrid composites using the hand lay-up method, followed by compression molding process [7, 8]. These fillers are selected because they can withstand high temperatures, form thin transfer films, and have low thermal expansion. The epoxy resin (LAPOX L-12) and hardener K-6 are used as matrix and hardener. The fillers are thoroughly mixed in epoxy with hardener before being poured on the glass fiber fabric in the mold. The mixing ratios of compositions are presented in Table 3.1. The composite plate is prepared with dimension of 250 mm × 250 mm × 3 mm. A diamond tipped cutter is used to cut the wear test samples from cured laminated hybrid composites to about 6 mm × 6 mm × 3 mm size. The dry sliding wear tests are performed using a POD machine. The counterface material is alloy steel of 8 mm thickness and 165 mm diameter. The surface roughness of the counterface is 1.2 μm. A wear track diameter of 130 mm is used to perform the tribo behavior tests. Before the test, the sample surface is rubbed over a 600 grade SiC sheet.

The effect of sliding velocities on wear rate for various compositions of glass fiber reinforced epoxy composites is shown in Figure 3.8. Increasing the sliding

**Table 3.1** Composition details for composite preparation.

| | | | Fillers | |
|---|---|---|---|---|
| Specimen code | Epoxy volume (%) | Glass fiber volume (%) | Silicon carbide | Graphite |
| A | 50 | 50 | — | — |
| B | 40 | 50 | 5 | 5 |
| C | 35 | 50 | 10 | 5 |

Basavarajappa and Ellangovan 2012 [7]. Reproduced with permission from Elsevier.

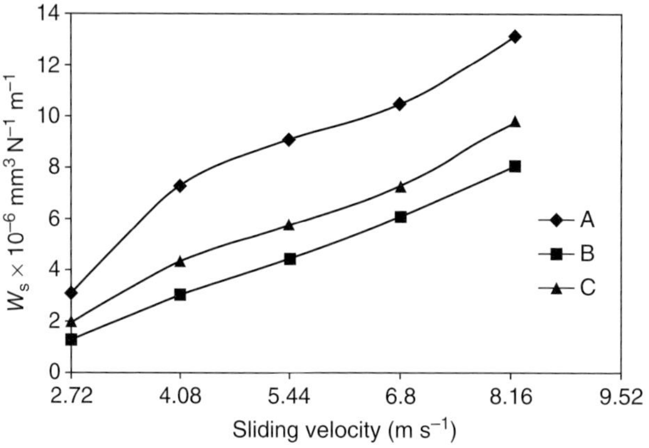


**Figure 3.8** Specific wear rate against various sliding velocity of fillers and glass fiber reinforced epoxy composites (testing parameter: 60 N and 3000 m). *Source:* Basavarajappa and Ellangovan 2012 [7]. Reproduced with permission from Elsevier.

distance increases the wear rate while incorporation of fillers in composites reduces the wear rate. Plowing and asperities are the major actions on the worn surfaces. This could affect the wear behavior. In experiments, the poor conductivity of epoxy can increase the interface temperature at higher sliding speeds. This temperature could melt the polymer and form a molten thin film on the interface. It can affect the interfacial bonding between the fiber and the matrix at the junction and reduce the wear rate as well as propagation of crack on the subsurface. The higher thermal conductivity and thermal expansion properties of fillers can have superior wear resistance compared to pure glass and epoxy, as shown in Table 3.2. Incorporating the lower thermal conductivity filler up to 10% can increase the rate of wear slightly when compared to equal percentage of fillers reinforcement.

Initially, the specific wear rate is found to be high for all composites at 20 N (Figure 3.9). If the applied load is increased up to 40 N, the specific wear rate falls because of change in wear process. Then the minimum wear rate occurs at 60 N and it increases again when the applied load is increased. The specific wear rate increases owing to the adhesive wear mechanism occurring on the counterface when more epoxy comes in contact during initial running. Subsequently, an equal amount of fillers in the composites shows low wear rate because both fillers are hard to wear simultaneously. Figure 3.10 shows a plot of the sliding distance against wear rate. The unfilled glass fiber epoxy composite shows higher wear rate than the fillers reinforced composites. The formation of air bubbles

**Table 3.2** Thermal property of the composition.

| Materials | Thermal conductivity ($W\,m^{-1}\,K^{-1}$) | Thermal expansion ($\times 10^{-6}\,K^{-1}$) |
|---|---|---|
| Silicon carbide | 3.8 | 7.6 |
| Graphite | 25–470 | 1.2–8.2 |
| Epoxy | ~0.45 | — |

Basavarajappa and Ellangovan 2012 [7]. Reproduced with permission from Elsevier.

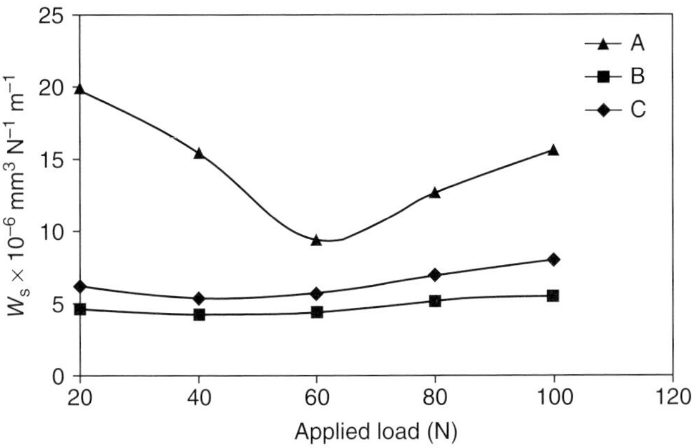


**Figure 3.9** Specific wear rate versus various applied load of fillers and glass fiber reinforced epoxy composites (testing parameter: 5.44 m s$^{-1}$ and at 3000 m). *Source:* Basavarajappa and Ellangovan 2012 [7]. Reproduced with permission from Elsevier.

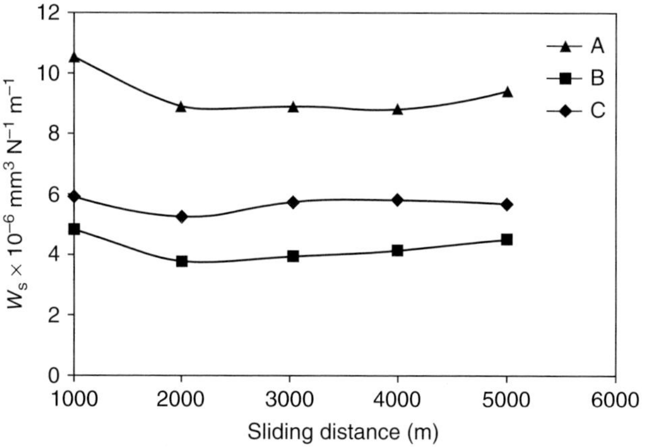


**Figure 3.10** Specific wear rate against various sliding distances of fillers and glass fiber reinforced epoxy composites (testing parameter: 5.44 m s$^{-1}$ and at 60 N). *Source:* Basavarajappa and Ellangovan 2012 [7]. Reproduced with permission from Elsevier.

and voids is unavoidable in unfilled composites because a viscous flow of matrix occurs in sliding. The combined effects of fillers in composites can reduce the voids and also increase the wear resistance. The presence of SiC particle enhances the adhesion, which reduces the matrix wear. At higher loads, the epoxy is softened by releasing additional frictional heat, thereby loosening the SiC particles. These SiC particles and the formed graphite film are removed from the specimen surface layer, and this forms a thin transfer film on the counterface. It exhibits high lubricity in a collective manner. This film can reduce the frictional heat and the damage to epoxy, glass fibers, and their bonding.

Basavarajappa et al. [8] investigated the effect of graphite filler contents on weight loss in glass fiber reinforced epoxy composites, as shown in Figure 3.11. The initial weight of the pin was measured by using a digital electric balance machine before tribo tests. The least count of the machine was 0.0001 g. At the end of the experiment, the pin was again weighed in the machine and differences between the two weights were taken as weight loss of the pin, and then the outcomes were plotted as weight loss against sliding distance and applied load against sliding distance. Figure 3.11a shows the weight loss of unfilled and filled glass fiber composites at a constant applied load of 60 N and sliding distance of 3000 m. It is clear that weight loss decreases with increase in graphite, but it increases with increase in the sliding velocity from 2.27 to 8.16 m s$^{-1}$. The fiber and matrix wear debris are well spread on the counterface up to a sliding distance of 6.8 m s$^{-1}$ due to a low degree of fiber breakage. If the sliding distance is increased to 8.16 m s$^{-1}$, the matrix face has greater wear and a high degree of fiber breakage is observed, leading to more wear debris formation on the counterface. Similar trends are also observed on varying the applied load (Figure 3.11b). Weight loss slowly goes up till 60 N for both 5% and 10% filled glass–epoxy composites, but identical weight loss is found for both filled composites. With further increases in load up to 100 N, the highest weight loss is observed for all composites. Applied load up to 80 N shows good spread of debris due to matrix wear, and

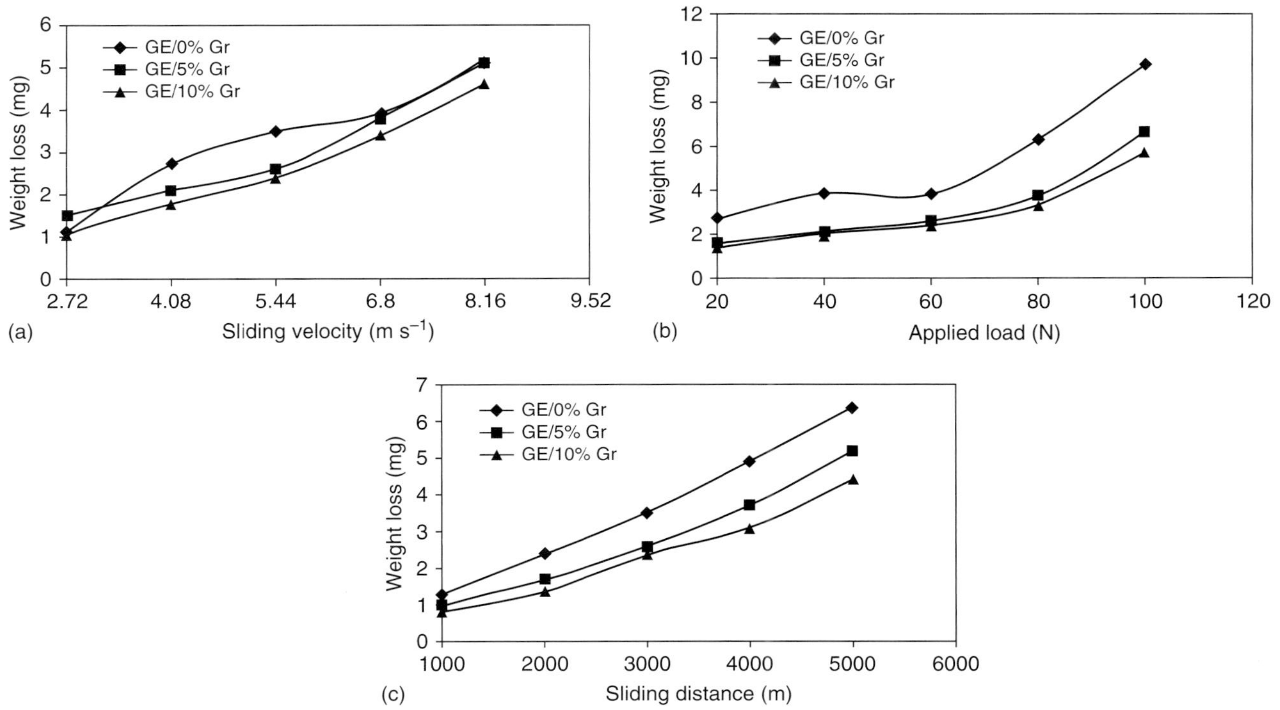


**Figure 3.11** Weight losses versus wear testing parameters: sliding velocity, applied load, and sliding distance. *Source:* Basavarajappa et al. 2009 [8]. Reproduced with permission from Elsevier.

also less wear debris is formed on the counterface. It affects the change of wear. At applied load 100 N, a few discontinuous glass fibers and void formation are seen on the surface of the pin. This could increase the weight loss. Weight loss increases linearly with sliding distance (Figure 3.11c). Extreme weight loss is found for the composite containing 10% graphite.

Figure 3.12 shows the effect of graphite filler on wear rate and $\mu$ of glass fiber fabric reinforced epoxy composites at constant sliding distance (1.2 km) and disk velocity ($5\,m\,s^{-1}$) [9]. Perhaps the applied loads and graphite weight contents have synchronized effects on specific wear rate and $\mu$ of the composites. Increasing the applied load increases the wear rate due to more graphite fillers coming out from the surface. However, a lower wear rate is found for composites containing 3% of graphite. Maximum wear rate is observed for 9% graphite composites at 90 N load. Also, the COF increases with increasing the graphite filler up to 6% and then a decreasing trend is observed at 9% graphite filler. At higher loads, the formation of thin film on the counterface can act as lubricant and it reduces the COF for all composites. The wear morphology of 3% graphite containing glass fiber composites at different applied loads is shown in Figure 3.13. It is clear that the wear loss of 3% graphite loaded composites is found to be less at 30 N when compared to higher loads. This is proved by the worn surface

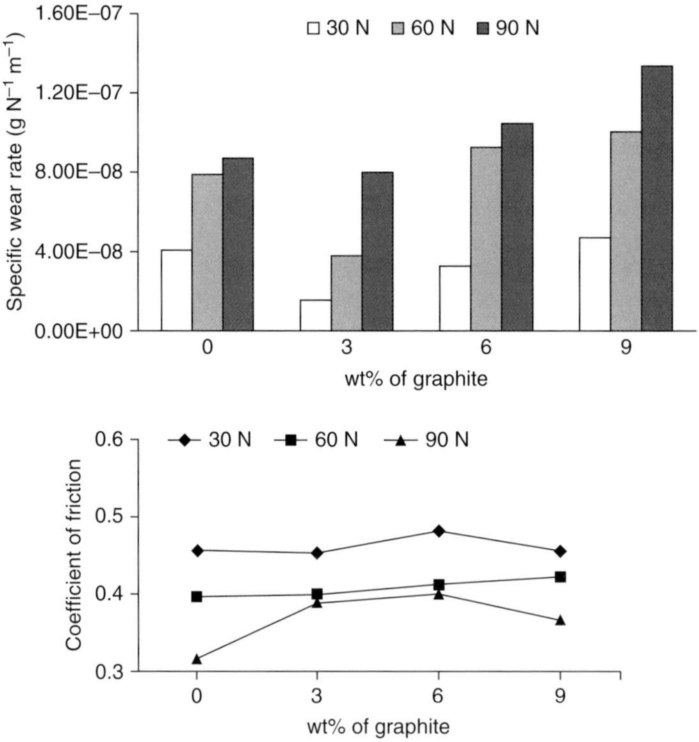


**Figure 3.12** Specific wear rates and coefficient of friction of pure and graphite filled glass fabric reinforced epoxy composites (normal applied loads 30, 60, and 90 N). *Source:* Shivamurthy et al. 2013 [9]. Reproduced with permission from Elsevier.

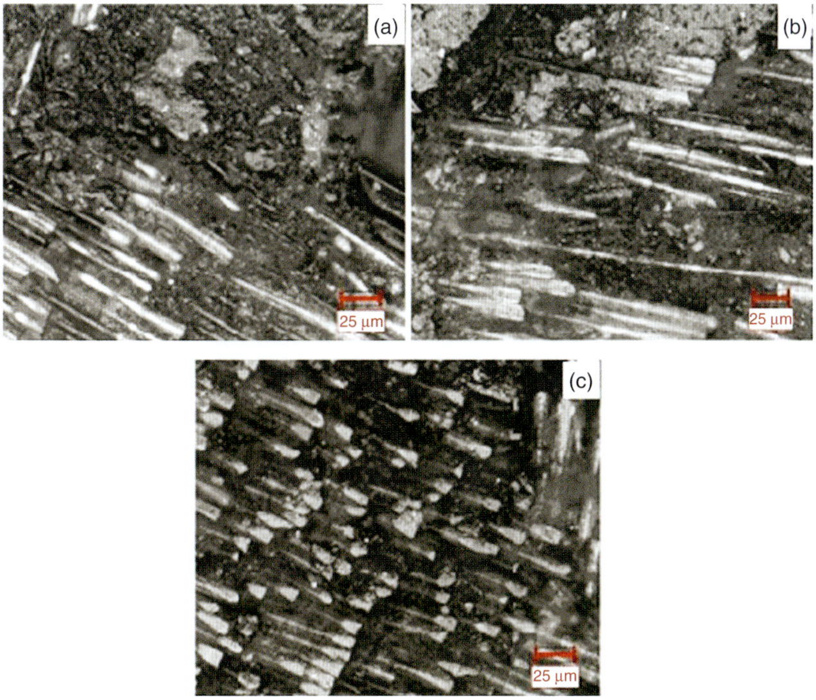


**Figure 3.13** Worn surface morphology of 3% graphite filler containing glass fiber composites at applied loads of (a) 30 N, (b) 60 N, and (c) 90 N. *Source:* Shivamurthy et al. 2013 [9]. Reproduced with permission from Elsevier.

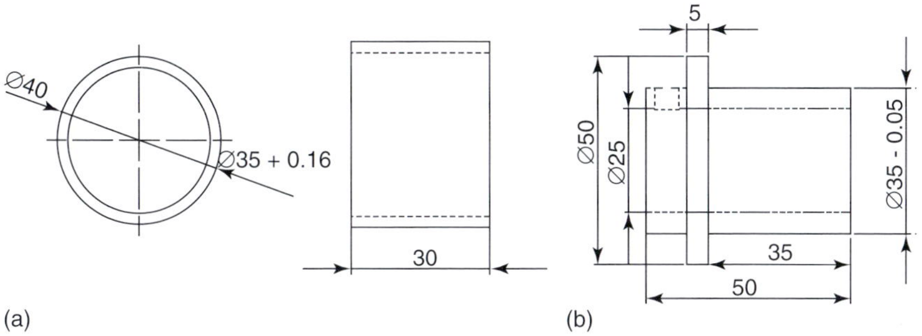


**Figure 3.14** (a,b) Dimensions of the bearing and the counterpart.

morphology; Figure 3.13a shows that part of the glass fiber appears on the worn surface. But at a higher load of 90 N, the surface damage is observed in two areas, fiber–matrix interface and matrix regions. Therefore, heavy damage occurs on the glass fiber as shown in Figure 3.13c.

### 3.4.3 Case Study

PA66 and glass fiber are used to prepare PTFE polymer journal bearing [10]. The dimensions of the counterpart and journal bearing are given in

Figure 3.14. The counterpart is made of DIN 52100 steel and is mounted on the shaft in the test apparatus. The LPV journal bearing apparatus is used to measure the COF and surface temperature against sliding distance. Two counterparts are prepared to minimize the heating effect on the surface. The specimens are cleaned before and after test by methyl alcohol and they are dried in an oven at 80 °C for 1 h. The PTFE bearing is prepared by three different compositions, PA66, PA66/18% PTFE, and PA66/20% glass fiber/25% PTFE. These are fitted in the polymer journal bearing montage shown in Figure 3.15.

Figures 3.16 and 3.17 show plots of the friction coefficients and contact temperature against sliding distance of all specimens. The friction coefficients and contact temperatures of all specimens exhibit increase on increasing the sliding distance and velocity. During experiments, the accumulated heat on PA66 and PA66 + 18% PTFE specimen surfaces is not distributed. Then the friction coefficients of these specimens are higher when compared to the PA66 + 20% GFR + 25% PTFE journal bearing. The glass fiber journal bearing has high thermal conductivity, which makes for better heat transmittance and decreases the friction coefficients. The PTFE has poor thermal conductivity; use of glass fibers in PTFE composite improves the properties by incorporation of higher thermal conductivity materials. The PTFE prevents the plow out of glass fiber on the counterface, which forms a film layer. As result, the glass fibers in journal bearing show better wear behaviors than others.

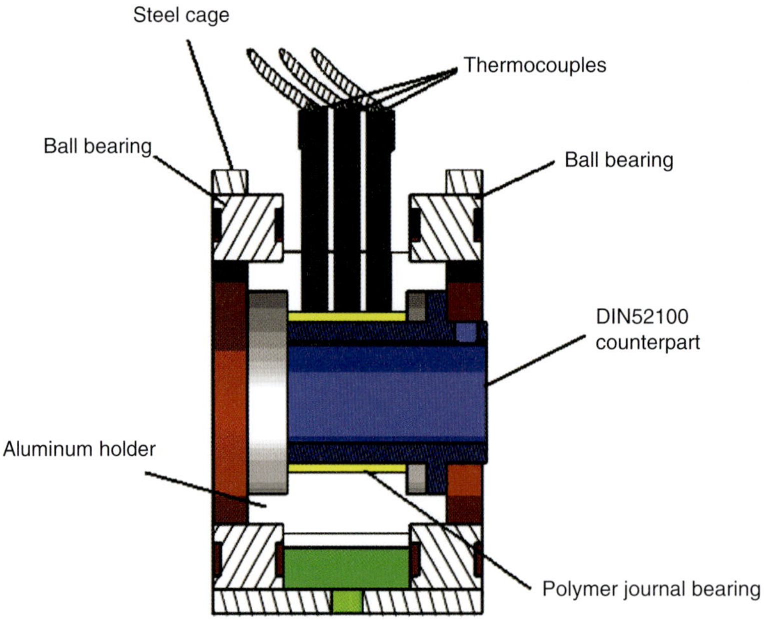


**Figure 3.15** Polymer journal bearing montage.

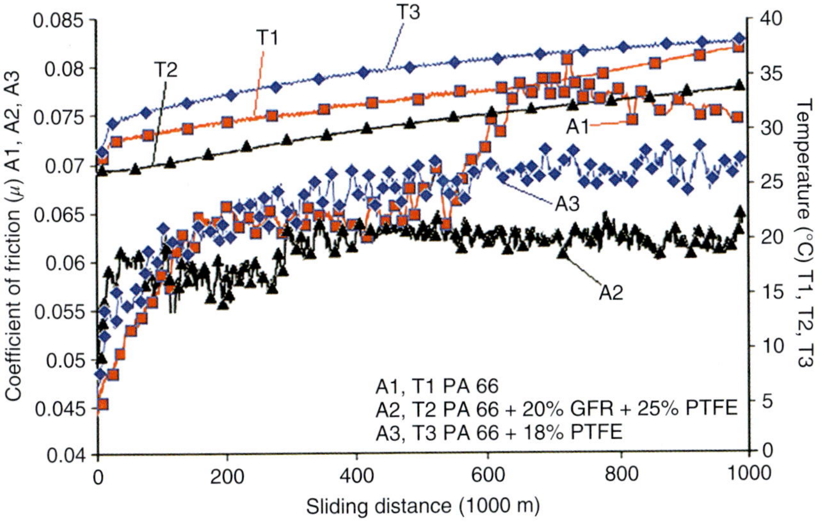


**Figure 3.16** Plot of the friction coefficient and contact temperature versus sliding distance of journal bearing at 0.5 m s$^{-1}$ and 0.238 MPa.

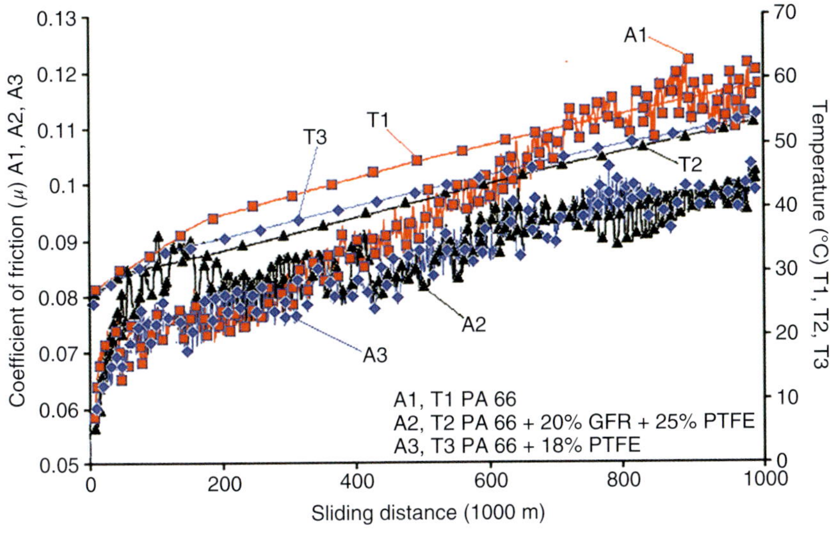


**Figure 3.17** Plot of the friction coefficient and contact temperature versus sliding distance of journal bearing at 1 m s$^{-1}$ and 0.238 MPa.

## 3.5 Conclusion

The influence of various fillers with glass fiber reinforced polymer composites on wear behavior has been extensively discussed. POD and BOD experiments were used to calculate the weight loss, specific wear rate, and friction coefficient by setting various testing parameters. The parameters varied were applied normal load, sliding velocity, and sliding distance. Addition of fillers in the glass fiber composites heavily affected the wear behaviors. Wear and weight loss increased with increase in the applied load, velocity and distance. But incorporation of fillers in glass–epoxy composites lowered the wear characteristics. This may be due to formation of a thin transfer film and more wear debris on the counterface. So the graphite filler is a suitable material for the development of glass fiber reinforced polymer composites for various products.

## References

1 Sathishkumar, T.P., Satheeshkumar, S., and Naveen, J. (2014). Glass fiber-reinforced polymer composites – a review. *Journal of Reinforced Plastics and Composites* 33 (13): 1258–1275.
2 Nirmal, U., Hashim, J., and Megat Ahmad, M.M.H. (2015). A review on tribological performance of natural fibre polymeric composites. *Tribology International* l83: 77–104.
3 Nirmal, U., Low, K.O., and Hashim, J. (2012). On the effect of abrasiveness to process equipment using betel nut and glass fibres reinforced polyester composites. *Wear* 290–291: 32–40.
4 Thongsang, S., Vorakhan, W., Wimolmala, E., and Sombatsompop, N. (2012). Dynamic mechanical analysis and tribological properties of NR vulcanizates with fly ash/precipitated silica hybrid filler. *Tribology International* 53: 134–141.
5 Palabiyik, M. and Bahadur, S. (2002). Tribological studies of polyamide 6 and high-density polyethylene blends filled with PTFE and copper oxide and reinforced with short glass fibers. *Wear* 253: 369–376.
6 Li, D.-X., You, Y.-L., Deng, X. et al. (2013). Tribological properties of solid lubricants filled glass fiber reinforced polyamide6 composites. *Materials and Design* 46: 809–815.
7 Basavarajappa, S. and Ellangovan, S. (2012). Dry sliding wear characteristics of glass–epoxy composite filled with silicon carbide and graphite particles. *Wear* 296: 491–496.
8 Basavarajappa, S., Ellangovan, S., and Arun, K.V. (2009). Studies on dry sliding wear behaviour of graphite filled glass–epoxy composites. *Materials and Design* 30: 2670–2675.
9 Shivamurthy, B., Udaya Bhat, K., and Anandhan, S. (2013). Mechanical and sliding wear properties of multi-layered laminates from glass fabric/graphite/epoxy composites. *Materials and Design* 44: 136–143.
10 Demirci, M.T. and Duzcukoglu, H. (2014). Wear behaviors of Polytetrafluoroethylene and glass fiber reinforced polyamide 66 journal bearings. *Materials and Design* 57: 560–567.

# 4

# Tribological Characterization of Jute/Glass Hybrid Composites

*Narayanan Venkateshwaran, Aziz Alavudeen, and Vajjiram Santhanam*

*Rajalakshmi Engineering College, Department of Mechanical Engineering, Rajalakshmi Nagar, Thandalam, Chennai 602105, Tamil Nadu, India*
*Kalasalingam University, Department of Mechanical Engineering, Anand Nagar, Krishnankoil 626126, Tamil Nadu, India*

## 4.1 Introduction

The importance of wear behavior is crucial as most of the design applications involve moving parts or relative motion between them causing "wear and tear." It has been reported that wear contributes the maximum failure cost to the total cost of industries. Contributing factors such as friction, heat, and wear cannot be excluded; yet, they can be minimized [1, 2]. Friction and wear are two important aspects in tribology. They exist during the relative motion among the solid surfaces, and have a tendency to dissipate energy and weaken the material performance. Nowadays, polymers are also used as frictional components in many applications. In order to improve the tribological behavior of neat polymers, the most widely used method is reinforcing fibers in polymers to fabricate composites [3]. Familiar reinforcements for polymers are glass, aramid, and carbon fibers, which are commonly used in composites fabrication. Fiber reinforced polymer (FRP) composites have been established as one of the most capable modern materials to replace conventional metals and alloys in several structural and tribological applications in recent years. These FRP composites are being fabricated using thermoplastics and thermosets such as matrix and synthetic and natural fibers as reinforcements along with small amounts of inorganic and organic materials as fillers. Such FRPs possess excellent properties due to their high specific strength, desired properties, and resistance to wear and corrosion. Synthetic fibers, namely glass, carbon, and aramid, have been used extensively along with thermosets and thermoplastics to develop FRP materials [4]. The main factors that contribute to the superior performance of the FRP composite are fiber length, fiber fraction, and interphase adhesion between matrix and fibers. The addition of fibers in matrix plays a major role in achieving the desired strength, stiffness, and tribological properties of the composites [5].

*Synthesis and Tribological Applications of Hybrid Materials*, First Edition.
Edited by Mohammad Jawaid, Rajini Nagarajan, Jacob Sukumaran, and Patrick De Baets.

Over the past era, much work has been carried out to establish the connection between fiber length and mechanical performance of FRP composites. Many researchers [6–10] have proposed various models to predict the mechanical properties of the composites, namely stiffness, strength, and impact properties. Most of the research investigations specify that fiber length is a crucial parameter in defining the mechanical performance of the composites. In contrast, understanding the effect of fiber parameters on wear and friction of composites is still deficient, in comparison to other factors [11–14]. Abrasive wear is the most severe form of wear in numerous applications using bearings, earth moving equipments, mining, etc. [15]. In proper designing of FRPs, the interfacial adhesion between matrix and reinforcement plays a vital role in governing the properties of the composites. Friedrich [16] reported that continuous fiber reinforcement in composites is more effective in reducing abrasive wear, because in longer fibers damages are localized as wear against abrasive particles and the remaining fibers still contribute to the wear resistance of composites. It was also found that in erosion, the fiber length did not contribute significantly, especially at high impact angles or erosive particles. Glass fiber (GF) reinforced composite has high mechanical strength and excellent wear resistance. In addition, GFs are widely available and quite cheap. The carbon fibers reinforced with polymer have lower coefficient of friction and better wear resistance as compared to a matrix when tested against steel surfaces [17]. Owing to the environmental issues that arise in using glass fibers as reinforcement, researchers are in the process of identifying a substitute. The literature on using natural fibers as reinforcement shows that they have high specific mechanical properties, along with being a biodegradable alternative to glass fibers [18]. Regardless of the importance and environment-friendly nature of natural fibers, their application has been limited to structural applications because of the lower strength and stiffness compared with synthetic fiber based composites [19]. Also, there are some studies on the possibility of natural fibers replacing glass fiber reinforcement. Little information with reference to the tribological behavior of natural fiber composites [11, 20–23] has been reported as of now. Although stiffness and strength are the shortcomings of bio-composites, they can be overcome by proper structural configurations and better arrangement of laminate in specific locations for desired performance. Further, this can be addressed by the combination of natural fibers and synthetic fibers [24]. Hence, the current work aims to investigate the possibility of using jute/glass fiber as reinforcement in polyester resin for tribological applications. Hybrid composites of jute and glass fiber woven mat was fabricated using hand layup method as shown in Figure 4.1.

## 4.2 Materials and Method

The commercially available polyester resin with polyamine hardener was considered as the matrix material in this study. Jute fibers and glass fibers were used as reinforcements. Polyester was purchased from a local dealer in Chennai and jute

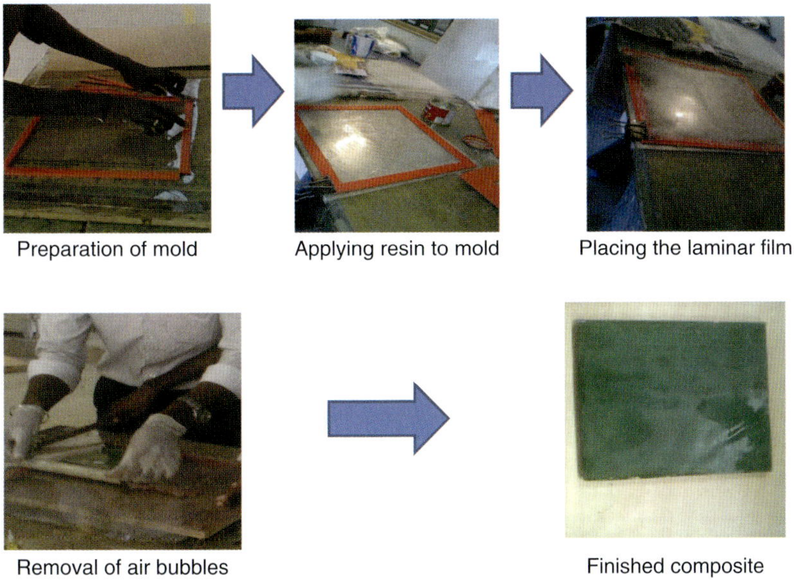


**Figure 4.1** Fabrication process.

fibers were obtained from the Jute Emporium, Chennai, and glass fibers were purchased from Sakthi Glass Fibers Private Ltd., Chennai. Hybrid laminates of jute and glass mat were prepared by hand lay-up technique. A wooden mold of dimension $(300 \times 300 \times 10)$ mm was used for casting the composite laminates. A gel coating was applied on the inner side of the mold, and mold release spray was used for easy removal of the composite fabricated. The hand lay-up technique was used to manufacture a composite sheet of 10 mm thickness at room temperature. The neat resin composite plate, also made with the above dimensions, was fabricated for comparison purpose. Figure 4.1 shows the fabrication process of the laminate and Figure 4.2 shows the ASTM standard specimen used

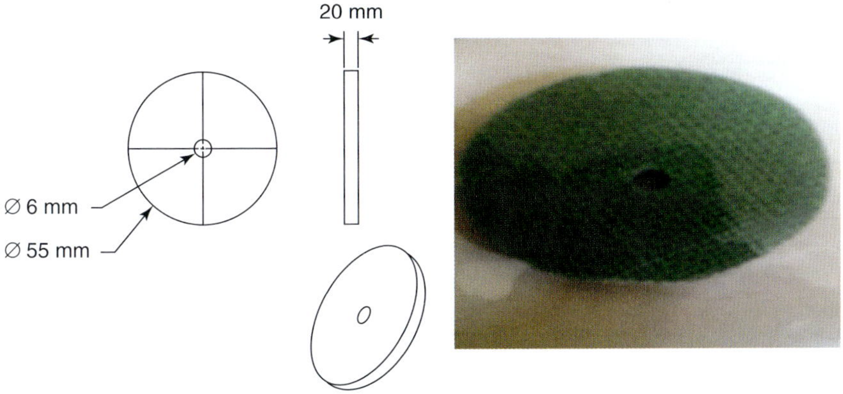


**Figure 4.2** Specimen used in this work.

**Table 4.1** Process parameters used in this work.

| Composite No | Load (N) | Sliding distances (m) |
|---|---|---|
| 1 | 5 | 300 |
| 2 | | 600 |
| 3 | | 1200 |
| 4 | | 1500 |
| 5 | 10 | 300 |
| 6 | | 600 |
| 7 | | 1200 |
| 8 | | 1500 |
| 9 | 15 | 300 |
| 10 | | 600 |
| 11 | | 1200 |
| 12 | | 1500 |

in this work. Wear test was performed using pin-on-disk (P-o-D) equipment according to ASTM D3702. Prior to the start of the experiment, surfaces of the samples and the disk were cleaned with a soft paper soaked in acetone and dried fully. The specimen was initially weighed to an accuracy of 0.001 g in an electronic balance machine. The differences between the initial and final weights were used to measure the sliding wear loss. Five numbers of samples were tested for each combination.

The frictional coefficient was calculated as a ratio between the tangential forces and the normal load. The mass losses of the sample were measured soon after the wear test to compute the specific wear rate using Equation (4.1):

$$W_s = \Delta m / \rho t V_S F_N \, \mathrm{mm^3 N^{-1} m^{-1}} \tag{4.1}$$

in which $F_N$ is the normal load applied on the specimen during sliding, $\Delta m$ is the specimen's mass loss, $\rho$ is the density of the specimen, $V_S$ is the velocity of the specimen, and $t$ is the time.

## 4.3 Results and Discussion

The load, sliding velocity, and sliding distance are the main parameters that influence the friction and wear loss of the components subjected to relative motion. The two foremost mechanisms that contribute to the friction between the composite disk and steel pin are adhesion and deformation. The relative contribution of these two depends on the load, and the chemical, mechanical, and geometrical properties. On smooth surfaces, adhesion becomes prominent whereas in rough surfaces, the deformation is important. The effects of the

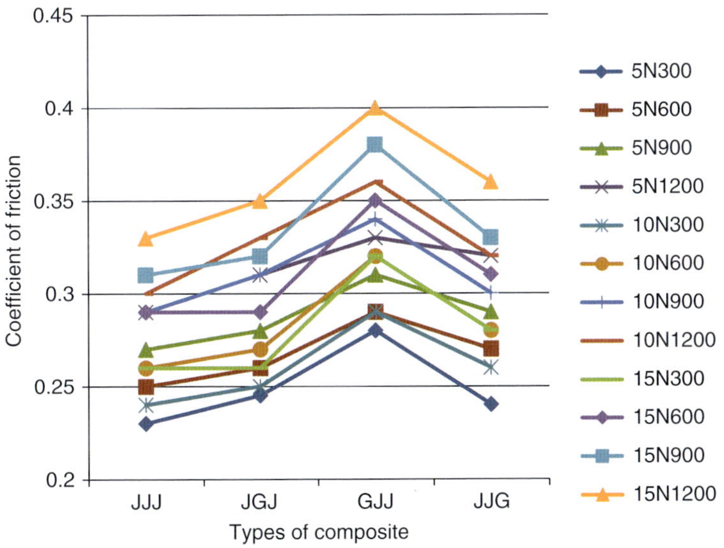


**Figure 4.3** Coefficient of friction of various composites.

normal load and the sliding distances over the coefficient of friction of various composite specimens are shown in Figure 4.3.

The coefficient of friction starts with a value of 0.23 for the jute fiber composite at 5 N load and 300 rpm and then it reaches a value of 0.33 at 15 N and 1200 rpm. With the addition of glass fiber as the middle layer in the lamina, the coefficient of friction starts from 0.245 and increases to 0.35 at 15 N and 1200 rpm. For the case of glass fiber placed as the first layer, the coefficient of friction ranges from 0.28 to 0.4 whereas for glass fiber placed as the third layer, the coefficient of friction ranges from 0.24 to 0.36. This shows that with the addition of glass fiber the coefficient of friction increases; this may be because when the glass fiber composite comes in contact with another moving surface, the surface heat generated is high when compared with the jute fiber composite. Further, during sliding, the frictional force caused might have pulled out the jute fiber along with wear debris. These dragged out jute fiber sticks on the surface of the pin might have acted as a three body rolling element, which evades the direct contact of the pin with the counter surface and hence results in reduction of coefficient of friction [25].

Figure 4.4 shows the influence of load and speed over the specific wear of various fabricated composites. The specific wear rate also increases with increase in the load and sliding speed. Of the entire composite fabricated, the JJJ composite has the lowest specific wear rate. The figure clearly shows that with the addition of glass fibers, the wear rate increases. This suggests that glass fiber addition increases the abrasive nature of the composite with the formation of wear debris particles at the interface, which in turn increases the friction coefficient and wear rate [26]. The addition of glass fibers increases the wear properties, which increases the fiber pullout and lessens adhesion between the fiber and the matrix [13].

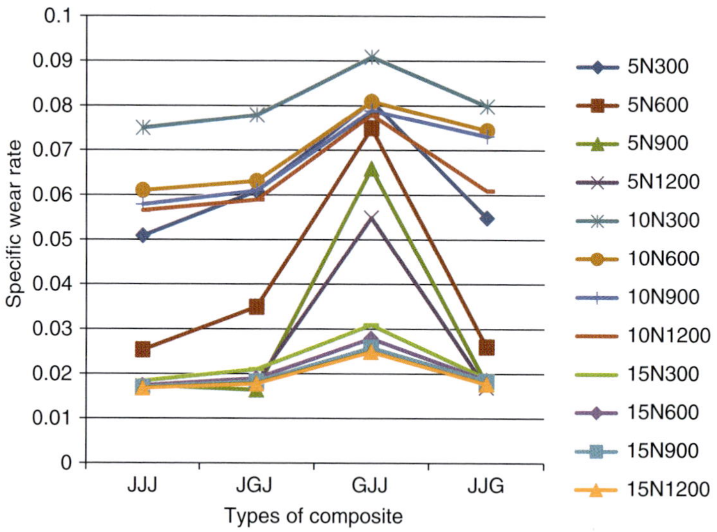


**Figure 4.4** Specific wear rate of various composites.

Figure 4.5 shows the influence of load and speed on the frictional force of the fabricated composites. It shows that the load applied has a significant impact on the frictional force rather than the speed of operation. It is understood that the relatively soft and non-abrasive nature of jute fiber has the ability to distort well under sliding conditions, making jute fiber a suitable reinforcement for low-load sliding conditions. In a dissimilar way, glass fibers are abrasive and brittle in nature, have less capability to deform, and are liable to cracking and debonding [27].

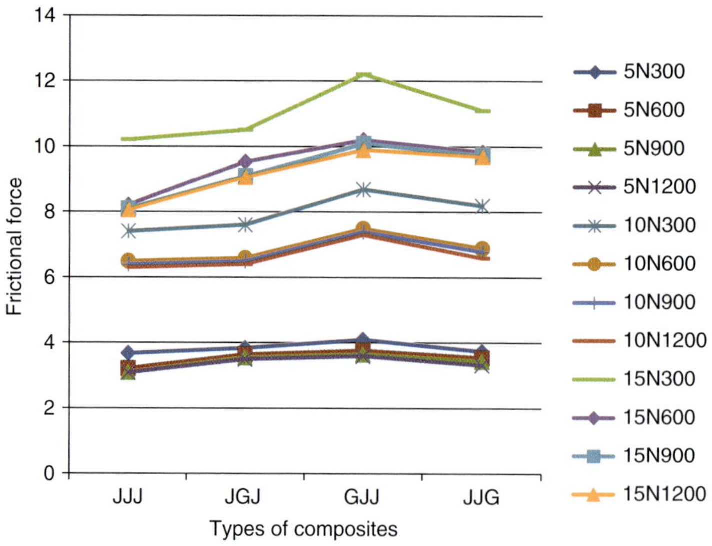


**Figure 4.5** Frictional force of various composites.

These exposed and deboned fibers rupture due to frictional load, providing a higher wear rate as seen from Figure 4.5.

## 4.4 Micrograph Analysis

To understand the outcome of parameters on the friction and wear characteristics of fabricated polymer and composite materials, the damaged surface of the specimens is examined using scanning electron microscopy (SEM) images. Figure 4.6 shows the irregularities developed over the surface of JJJ laminate at 5 N and 300 rpm. This shows that at low loads, the frictional heat generated is low, which in turn lowers the formation of polymer film over the pin surface. In addition, the SEM image shows the irregularities on the surface of the matrix due to the high bonding between matrix and reinforcement.

Figure 4.7 is the micrograph of JJJ composite at 15 N and 1200 rpm. It shows that at higher loads, areas of polymer film are formed due to severe plastic deformation and also, the worn surfaces show the formation of a fiber-rich surface [28]. The look of the smooth deformed layers made during sliding process endorses that at higher load, considerable plastic deformation of particles from the matrix takes place and settles as a shielding layer. These layer formations provide an even distribution of interface stresses of lower magnitude and thus they have lower wear [29].

Figure 4.8 shows the SEM image of three layers of glass fiber composite at 15 N and 1200 rpm. It shows poor bonding at the interface between the fiber and the matrix. It also shows that the mechanism of fiber removal in the woven mat is the complete removal of a layer itself. Since the complete removal of fibers takes place they are attached with the pin and this increases the coefficient of friction [19].

Figure 4.9 shows the SEM micrograph of GJJ composite at 15 N and 1200 rpm. It shows that at high load and speed the debris of matrix is formed. This formation

**Figure 4.6** SEM image of JJJ composite at 5 N and 300 rpm.

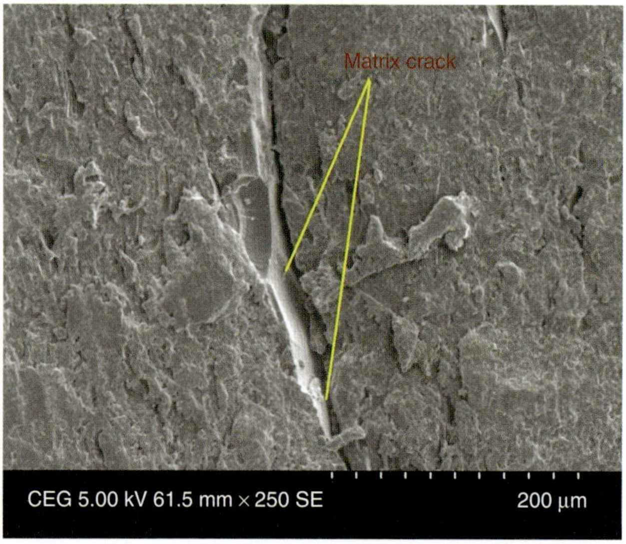


**Figure 4.7** SEM image of JJJ Composite at 15 N and 1200 rpm.

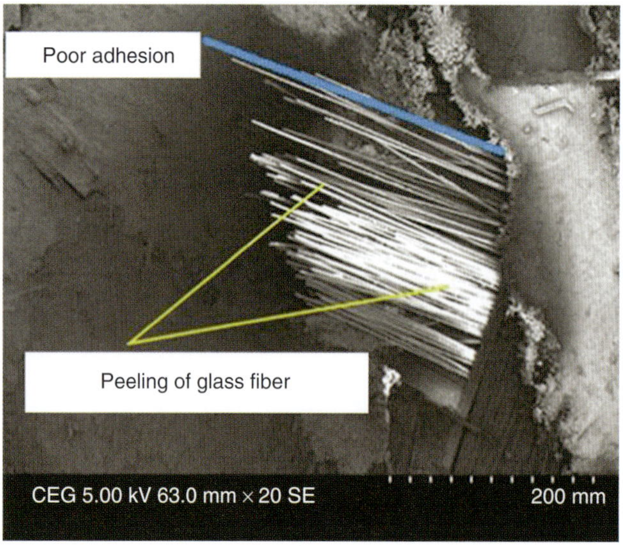


**Figure 4.8** SEM image of GGG composite at 15 N and 1200 rpm.

of debris exposes the fiber toward the pin and results in gradual transformation of adhesive wear to fatigue and slight abrasive wear [30].

Figure 4.10 shows the SEM images of JJG composite at 15 N and 1200 rpm. It shows the lack of formation of polymer film over the surface, confirming that the plastic deformation and adhesive or fatigue are responsible for the worn surface between the matrix and the counterpart due to repeated nature of stress [31]. At higher magnification, the micrograph (Figure 4.10b) displays widespread breakage and crack in the matrix at the interfacial regions along with wear trace.

Figure 4.9 SEM image of GJJ composite at 15 N and 1200 rpm.

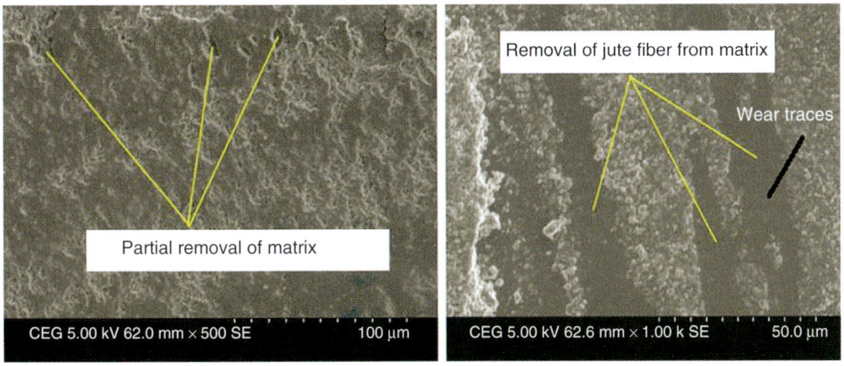


(a) (b)

Figure 4.10 (a,b) SEM image of JJG composite.

## 4.5 Conclusions

Tribological properties of various composites fabricated by varying the layer sequence and its performances were evaluated through P-o-D equipment. The results obtained are as follows:

1) The three-layer jute fiber composite possesses the lowermost coefficient of friction among the composites tested.
2) The addition of glass fiber increases the coefficient of friction with varying load and sliding distances. Hence, these composites can be suitable for friction pad, etc.

3) The addition of glass fiber also increases the specific wear of the composite. This suggests that glass fiber addition increases the abrasive nature of the composite with the formation of wear debris/particles at the interface.
4) The SEM micrographs of various composites are analyzed and show that the jute fiber added composite has adhesion wear. The addition of glass fiber into it changes the wear mechanism from adhesion to fatigue and slight abrasive wear.

## References


1 Omrani, E., Menezes, P.L., and Rohatgi, P.K. (2016). State of the art on tribological behavior of polymer matrix composites reinforced with natural fibers in the green materials world. *Engineering Science and Technology, an International Journal* 19: 717–736.
2 El-Sayed, A.A., El-Sherbiny, M.G., Abo-El-Ezz, A.S., and Aggag, G.A. (1995). Friction and wear properties of polymeric composite materials for bearing applications. *Wear* 184: 45–53.
3 Parikh, H.H. and Gohi, P.P. (2015). Tribology of fiber reinforced polymer matrix composites – a review. *Journal of Reinforced Plastics and Composites* 34 (16): 1340–1346.
4 Aldousiri, B., Shalwan, A., and Chin, C.W. (2013). A review on tribological behaviour of polymeric composites and future reinforcements. *Advances in Materials Science and Engineering* 10: 1–8.
5 Zsidai, L., De Baets, P., Samyn, P. et al. (2002). The tribological behaviour of engineering plastics during sliding friction investigated with small-scale specimens. *Wear* 253: 673–688.
6 Srivastava, V.K. and Pathak, J.P. (1996). Friction and wear properties of bushing bearing of graphite filled short glass fiber composite in dry sliding. *Wear* 197: 145–150.
7 Unal, H., Sen, U., and Mimaroglu, A. (2005). Abrasive wear behaviour of polymeric materials. *Materials and Design* 26: 705–710.
8 Hong, U.S., Jung, S.L., and Cho, K.H. (2009). Wear mechanism of multiphase friction materials with different phenolic resin matrices. *Wear* 266: 739–744.
9 El-Tayeb, N.S.M. (2008). A study on the potential of sugarcane fibers/polyester composite for tribological applications. *Wear* 265: 223–235.
10 Zhao, G., Ding, Q., and Wang, Q. (2016). Comparative study on the Tribological properties of the polyimide composites reinforced with different fibers. *Polymer Composites* 37: 2541–2548.
11 Chand, N. and Dwivedi, U.K. (2006). Effect of coupling agent on abrasive wear behaviour of chopped jute fibre-reinforced polypropylene composites. *Wear* 261: 1057–1063.
12 Zhang, H., Zhang, Z., and Friedrich, K. (2007). Effect of fiber length on the wear resistance of short carbon fiber reinforced epoxy composites. *Composites Science and Technology* 67: 222–230.
13 Chang, L., Zhang, Z., Breidt, C., and Friedrich, K. (2005). Tribological properties of epoxy nanocomposites: I. Enhancement of the wear resistance by nano-$TiO_2$ particles. *Wear* 258: 141–148.

14 Tiwari, S. and Bijwe, J. (2014). Influence of fiber–matrix interface on abrasive wear performance of polymer composites. *Journal of Reinforced Plastics and Composites* 33: 115–126.
15 Tiwari, S., Bijwe, J., and Panier, S. (2011). Role of nano YbF3 treated carbon fabric on improving abrasive wear performance of polyetherimide composites. *Tribology Letters* 42 (3): 293–300.
16 Friedrich, K., Zhang, Z., and Schlar, A.K. (2005). Effects of various fillers on the sliding wear of polymer composites. *Composites Science and Technology* 65: 2329–2343.
17 Bahadur, S. and Polineni, V.K. (1996). Tribological studies of glass fabric reinforced polyamide composites filled with CuO and PTFE. *Wear* 200: 95–104.
18 Pihtili, H. (2009). An experimental investigation of wear of glass fibre–epoxy resin and glass fibre–polyester resin composite materials. *European Polymer Journal* 45: 149–154.
19 Kishore, S.P., Seetharamu, S., Murali, A., and Kumar, R.K. (2001). On the SEM features of glass–epoxy composite system subjected to dry sliding wear. *Wear* 247: 208–213.
20 Dwivedi, U.K. and Chand, N. (2009). Influence of MA-g-PP on abrasive wear behaviour of chopped sisal fiber reinforced polypropylene composites. *Journal of Materials Processing Technology* 209: 5371–5375.
21 Bajpai, P.K., Singh, I., and Madaan, J. (2013). Tribological behavior of natural fiber reinforced PLA composites. *Wear* 297: 829–840.
22 Punyapriya, M. and Acharya, S.K. (2010). Anisotropy abrasive wear behaviour of bagasse fiber reinforced polymer composite. *International Journal of Engineering Science and Technology* 2: 104–112.
23 Yousif, B.F. and NSM, E.-T. (2009). Wet adhesive wear characteristics of untreated oil palm fiber reinforced polyester and treated oil palm fiber reinforced polyester composites using the pin on disc and block on ring techniques. *Proceedings of the Institution of Mechanical Engineers, Part J: Journal of Engineering Tribology* 224: 123–131.
24 Olea-Mejia, O., Brostow, W., and Buchman, E. (2010). Wear resistance and wear mechanisms in polymer + metal composites. *Journal of Nanoscience and Nanotechnology* 10: 8524–8530.
25 Jawahar, P., Gnanamoorthy, R., and Balasubramanian, M. (2006). Tribological behaviour of clay – thermoset polyester nanocomposites. *Wear* 261: 835–840.
26 Bharadwaj, R.K., Mehrabi, A.R., Hamilton, C. et al. (2002). Structure–property relationships in cross-linked polyester–clay nano composites. *Polymer* 43: 3699–3705.
27 Burris, D.L., Boesl, B., Bourne, G.R., and Gregory Sawyer, W. (2007). Polymeric nanocomposites for tribological applications. *Macromolecular Materials and Engineering* 292: 387–402.
28 Pihtili, H. and Tosun, N. (2002). Investigation of the wear behaviour of a glass-fibre-reinforced composite and plain polyester resin. *Composites Science and Technology* 62 (3): 367–370.
29 Hashmi, S.A.R., Dwivedi, U.K., Naik, A., and Chand, N. (2002). Effect of chemical treatments on abrasive wear behaviour of PP based lining materials. *Metals Materials and Processes* 14 (3): 235–240.

30 Deo, C. and Acharya, S.K. (2010). Effect of fiber content on abrasive wear of Lantana Camara fiber reinforced polymer matrix composite. *Indian Journal of Engineering and Materials Sciences* 17: 219–223.
31 Wong, K.J., Yousif, B.F., and Low, K.O. (2010). The effect of alkali treatment on the interfacial adhesion of bamboo fiber. *Proceedings of the Institution of Mechanical Engineers, Part L: Journal of Materials: Design and Applications* 224: 139–148.

# 5

# Glass Fiber Hybrid Effects in Assessing the Abrasive Wear Mechanisms of Naturally Woven Fabric/Polymer Composites Under Dry Conditions

*Irulappasamy Siva[1], Manoharan Thirukumaran[1], Jebas T.W. Jappes[1], Bhimappa Suresha[2], Subramaniyan A. Kumar[3], and Buradagunta R. Sunil[4]*

[1] *Kalasalingam Academy of Research and Education, Centre for Composite Materials, Department of Mechanical Engineering, Krishnankoil 626126, Tamil Nadu, India*
[2] *The National Institute of Engineering, Department of Mechanical Engineering, Mysore 570008, Karnataka, India*
[3] *Indian Institute of Technology Jammu, Jammu 181121, Jammu and Kashmir, India*
[4] *Rajiv Gandhi University of Knowledge Technologies (IIIT-AP), Department of Mechanical Engineering, Mylavaram Road, Nuzvid 521202, Andhra Pradesh, India*

## 5.1 Introduction

Natural fibers are now considered as suitable alternatives to synthetic fibers such as E-glass and other commercial fibers [1]. The major advantages of the natural fibers are that they are biodegradable, renewable, harmless during processing, etc. Hence, many automobile sectors have now started utilizing vegetable fibers as replacements to the synthetic fibers in use [2]. On the other hand, natural fibers have severe disadvantages such as swelling, high brittleness due to rich cellulose content, and unstableness to chemicals [3]. Hence, in order to attain and maintain the service standard, natural fiber composites need hybridization to provide additional support [4]. Hybridization refers to the inclusion of two or more reinforcements to a single matrix medium [5]. Relative weight percentage of the reinforcement eventually alters the properties of the final hybrid composites [6].

Coconut sheath is one of the less explored natural fibers [7] in the literature. The strength and properties of the coconut sheath are completely different from those of the so-called coir fiber [8]. Siva et al. [9] explored the dry sliding wear of the coconut sheath polyester composite; the authors have also discussed the effect of glass fiber hybridization on the dry sliding wear of the composite. Siva and Winowlin [10] studied the effect of silane coupling agent on the mechanical properties of coconut sheath polyester composites. The authors proved the influence of the silane coupling agent's coating on the coconut sheath fiber using scanning electron microscopy. Harish et al. [11] reported the advantages of "mat" type fibers compared to E-glass fibers.

*Synthesis and Tribological Applications of Hybrid Materials*, First Edition.
Edited by Mohammad Jawaid, Rajini Nagarajan, Jacob Sukumaran, and Patrick De Baets.

Abrasive wear occurs when a solid object is loaded and slid against particles of another material with a similar or higher roughness [12]. Its wear mechanism can be two-body abrasion, three-body abrasion, or both, depending on the wear debris removed from the contacting surfaces. Although abrasive wear involving fiber polymeric composites covers a huge area in the literature, the wear process particularly involving polymeric composites is still not well understood [13, 14]. This motivates the current work to be focused on abrasive wear behavior.

Nonetheless, substantial research has been done on the tribological properties of manmade fiber composites; very few reports are found with coconut sheath/polymer composites hybridized with glass fibers. The aim of this research is to study the possibility of using naturally woven coconut sheath fibers (which is seen in the hidden portion of the coconut branches) and glass fibers as reinforcements for tribological applications. Polyester composites based on woven coconut sheath and glass fibers (chopped strand mat) were fabricated, and their mechanical and abrasive wear behavior was explored eventually to analyze the hybridization effect of the fibers on its abrasive wear. Further, hardness and tensile strength of these composite materials are also discussed.

## 5.2 Experimental Details

### 5.2.1 Fiber and Matrix

Unsaturated polyester (USP) (general purpose grade: SBA2303) was used as the matrix. Reagents such as methyl ethyl ketone peroxide (MEKP) as catalyst and cobalt-naphthenate as accelerator were used. The resin and reagents were supplied by Vasavibala Resins, Chennai, Tamil Nadu, India. Trichlorovinyl silane from Sigma-Aldrich, Bangalore, Karnataka, India, was used as the coupling agent. Naturally woven coconut sheath was acquired from the nearest agriculture field. Commercially available E-glass chopped strand mat was also used as reinforcement for comparison purpose. Photographs of the two fibers used in this work are shown in Figure 5.1.

### 5.2.2 Fiber Surface Treatment

Naturally woven coconut sheath was initially battered with cold water and then dried. The dried fibers were immersed in 1 N alkali solution for an hour. After

**Figure 5.1** Photographs of coconut sheath (a) and glass fiber (b) used in this study.

a thorough wash using distilled water, the fibers were dried in a hot air oven at 60 °C for 4 h. Trichlorovinyl silane solution was prepared by diluting 0.3% trichlorovinyl silane in 1 l acidified water with a pH of 3.5. The dried fibers were then immersed into the silane solution for an hour followed by simple drying process in atmospheric condition for 24 h [15]. Dried fibers were cut into the required dimension and kept ready for molding.

### 5.2.3 Composite Fabrication

Compression molding technique was used in the fabrication of pure and hybrid composites. Preprocessed naturally woven coconut sheath and glass fibers were cut to the dimension of $180 \times 150\,mm^2$. Close mold was used in the compression molding process, and the mold cavity was protected with a releasing agent.

Referring to Table 5.1, the designations N and G represent two layers of coconut sheath and chopped strand glass mat, respectively. Catalyzed resin was coated on the bottom of the mold cavity followed by the stacking of six (overall optimized number) natural fiber and/or glass fiber laminas. For hybrid composites, the lamina numbers were relative. An optimized compression pressure of $50\,kg\,cm^{-2}$ was applied for composite molding. After 12 h of curing, the specimens were taken out from the mold and cut into the dimensions according to the respective ASTM standards. In the same manner, the glass fiber reinforced polyester composites were also prepared. For hybrid composites, intermediate positioning of the fibers was done.

### 5.2.4 Mechanical Testing

Tensile properties (ASTM D3039) of the coconut sheath and glass fiber reinforced USP composites and hybrid composites were measured at room temperature using an Instron (Series-3382) Universal Testing Machine. The tests were conducted at a crosshead speed of $2\,mm\,min^{-1}$. Some of the measured mechanical properties of composites are listed in Table 5.2. Shore hardness of composite samples was also measured using a Shore D hardness tester (Time Group Inc., Shore D Durometer TH210). Ten readings at different points were taken and the average values are reported in Table 5.2.

**Table 5.1** Formulation of composites.

| Composites | Specimen code | Polyester (wt%) | Coconut sheath (wt%) | Glass fiber (wt%) |
|---|---|---|---|---|
| Coconut sheath-USP | NNN/USP | 52 | 48 | — |
| Coconut sheath + glass fiber USP | NNG/USP | 52 | 31 | 17 |
| Coconut sheath + glass fiber USP | NGG/USP | 52 | 16 | 32 |
| Glass fiber-USP | GGG-USP | 52 | — | 48 |

**Table 5.2** Mechanical properties of composites.

| Property | NNN/USP | NNG/USP | NGG/USP | GGG/USP |
|---|---|---|---|---|
| Tensile strength, $\sigma$ (MPa) | 60 | 120 | 80 | 95 |
| Elongation (%) | 2.3 | 4.8 | 4.02 | 5.67 |
| Young's modulus, $E$ (GPa) | 2.608 | 2.5 | 1.990 | 1.676 |
| Density ($g\,cm^{-3}$) | 1.059 | 1.079 | 1.302 | 1.513 |
| Hardness (Shore D) | 75 | 88 | 87 | 78 |

### 5.2.5 Abrasive Wear Test

The modified dry sand silica/rubber wheel abrasion test setup (make: Magnum Engineers, Bangalore/Karnataka, India) as per ASTM G-65 was used to conduct three-body abrasive wear experiments and the schematic diagram is shown in Figure 5.2. In the present study, abrasive particles of AFS 60 grade silica sand were jagged in shape with sharp edges and were used as abrasive. The abrasive was fed at the contacting face between the rotating rubber wheel and the test sample. The test was conducted at the rotational speed of 200 rpm. The feed rate of the abrasive was $255 \pm 5\,g\,min^{-1}$. The sample was cleaned with acetone and dried. Its initial weight was determined in a high precision digital balance (0.1 mg accuracy, Mettler Toledo make) before it was mounted on the sample holder.

The abrasive was introduced between the test specimen and the rotating abrasive wheel composed of chlorobutyl rubber (hardness: Durometer-A 58-62). The test specimen was pressed against the rotating wheel at a specified force by means of the lever arm while a controlled flow of abrasive abraded the test surface. The rotation of the abrasive wheel was such that its contacting face moved in the direction of sand flow. The pivot axis of the lever arm lay within a plane,

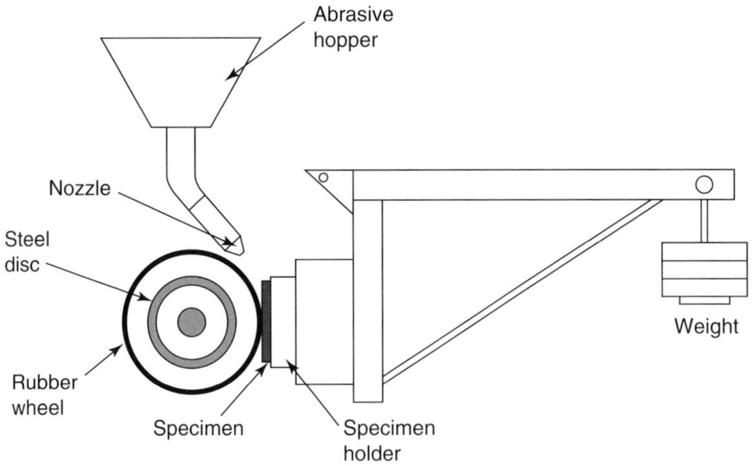


**Figure 5.2** Schematic diagram of dry sand rubber wheel abrasion tester.

which was approximately tangential to the rubber wheel surface and normal to the horizontal diameter along which the load was applied. At the end of the set test duration, the specimen was removed, thoroughly cleaned, and again weighed (final weight). The difference in weights before and after abrasion was determined. The experiment was carried out at two different normal loads of 24 and 36 N and at a constant sliding velocity of 2.15 $m\,s^{-1}$. Further, the abrading distances were 250, 500, 750, and 1000 m respectively.

The specific wear rate was calculated by the equation:

$$K_s = \frac{\Delta V}{L \times d} \mathrm{m^3 N^{-1} m^{-1}} \tag{5.1}$$

where, $\Delta V$ is the volume loss, $L$ is the applied load, and $d$ is the abrading distance.

## 5.3 Results and Discussion

### 5.3.1 Mechanical Properties

Table 5.2 represents the experimental results of the mechanically characterized virgin and hybrid composites of coconut sheath and glass fiber with polyester. Incorporation of a single glass fiber layer into NNN-USP composite resulted in significant enhancement in the mechanical properties. In the meantime, no such significant improvements were observed while reinforcing a single layer coconut sheath fiber into glass fiber composites. Such reverse hybridization weakens the structure, resulting in lower tensile strength at rupture. Hybridization, i.e. coconut sheath and glass fiber reinforcement into neat resin, increased the tensile strength of the neat polymer by 33% in the case of the single glass/natural fiber composite (NNG) and by 100% for single natural fiber/glass fiber composite (GGN). The hardness also increased from 75 to 88 for NNG and 78 to 87 for NGG composites, respectively.

The fractographs of the NNN-USP (Figure 5.3a) composites show poor adhesion with natural fiber and polyester matrix, and the presence of the hackles near

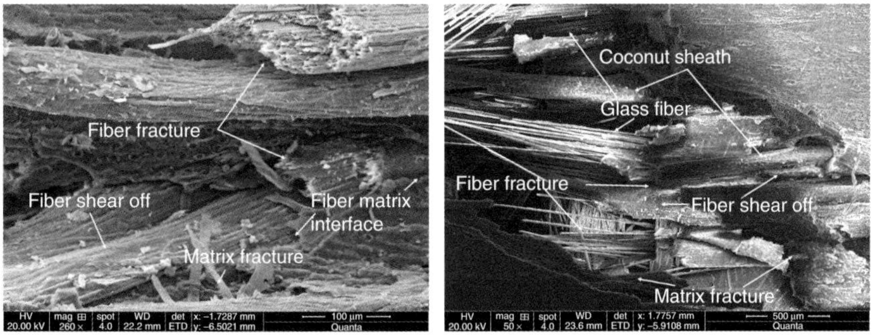


(a) (b)

**Figure 5.3** Fractography of NNN-USP composite (a) and NNG-USP composite (b).

the fractured matrix surface makes it evident that the newly identified coconut sheath fiber has the ability to be used as reinforcement to the polymer matrix. Similarly, in the fractography of the NNG-USP (Figure 5.3b), more fiber pull-outs and matrix failure were also observed. Hybridization imparts enhancement in the tensile strength of the coconut sheath reinforced composite. The newly identified coconut sheath (Figure 5.1a) has a mesh-like nature and the incorporation of glass fibers into the mesh of the fibers builds enhanced mechanical interlocking, which is attributed to improved tensile strength.

Increasing hybridization increased the tensile strength of the coconut sheath reinforced composite (Table 5.2). Pull-out fibers appeared with rich resin quantity, which shows better stress transfer from the matrix to the reinforcements.

### 5.3.2 Hardness (Shore-D) of Composites

Table 5.2 lists the Shore-D hardness of the composites for different stacking of coconut sheath and glass fiber layers in the USP matrix. It was seen that the value of Shore hardness of the composite increased from 75 to 88, when the coconut sheath was hybridized with glass fibers. This indicated that the glass fibers significantly increased the load-carrying capacity of coconut sheath polyester composite. The physical properties of natural fibers are mainly determined by their chemical and physical composition, such as structure of fibers, cellulose content, angle of fibrils, and cross section, and by the degree of polymerization [16]. The presence of glass fiber strengthens the coconut sheath polyester composite.

### 5.3.3 Specific Wear Rate

Fiber type and loading are known to improve many physical and/or mechanical and tribological properties of polymers [17]. Yousif et al. [18] investigated the wear and frictional behavior of epoxy composite based on treated betel nut fibers subjected to three-body abrasion using different abrasives/sliding velocities and at a constant applied load of 5 N. They revealed that the predominant wear mechanism of the treated betel nut fiber reinforced epoxy composite sliding against grain sands was plastic deformation, pitting and pull-out of betel nut fibers. The composite exhibited higher values of frictional coefficient when it was subjected to coarse sand. Higher weight loss is noticed at high sliding velocities. The specific wear rate for the composite subjected to three different sand particles follows the order coarse > grain > fine sands respectively.

A typical abrasive wear scar is illustrated in the Figure 5.4 for various applied loads and running distances. Figure 5.4a shows the wear scar of NNN-USP composite for 36 N applied load and abrading distance in steps of 250 m. One can see the increase in abrasive depth from 250 to 1000 m in the NNN-USP composites. Similarly for NGG-USP (Figure 5.4b) composites also an increased abrasive cavity is noted with the increase in abrading distance.

Figure 5.5 shows the wear volume as a function of abrading distance for the different applied loads. Figure 5.5a shows the effect of hybridization as a function of wear volume for 24 N applied load. An increasing trend in the wear volume

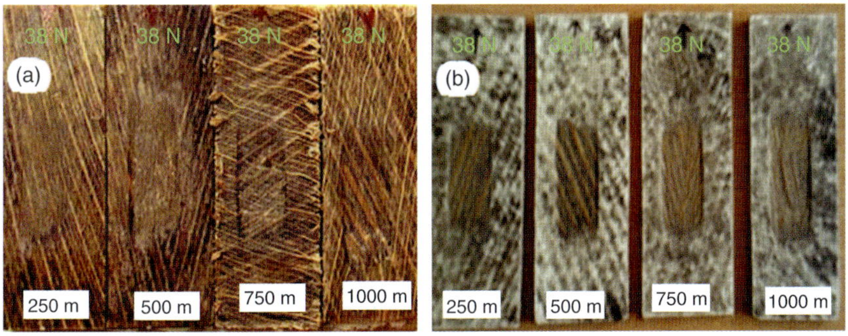


**Figure 5.4** Worn surface images of composites: coconut sheath composite (a) and hybrid composite (b).

was noted irrespective of the fiber type and purity in composite. Especially in the low loading condition, coconut sheath and glass fiber pure composites showed less mass loss compared to the hybrid composites. Among the pure composites, silane treated coconut sheath fiber reinforced polyester composite exhibits higher wear resistance (by means of less wear volume). In the meantime, inclusion of less glass fiber hybrid (NNG-USP) highly affects the wear resistance of the coconut sheath polyester composite. Further incorporation of more glass fibers slightly lowers the wear volume. Stacking the same group of fibers together into the composite promotes load bearability resulting less wear. Nevertheless, when abrading for a longer distance, this effect was suppressed. Figure 5.5b shows the wear volume of the pure and hybrid composites for 36 N applied load. Similar to the low loading results, at high loads only the coconut sheath polyester composites exhibit greater wear resistance. Incorporation of the glass fibers increased the wear volume of the composites. On the other hand, incorporation of the coconut sheath fiber into the glass fiber composites greatly reduced the wear volume of the glass fiber polyester composites.

Figure 5.6 shows the specific wear rate ($K_s$) as a function of abrading distance of various composite systems using silica sand as abrasive. Specific wear rate data reveal the fact that it tends to decrease with increasing abrading distance from 250 to 1000 m. However, the higher naturally woven coconut sheath (N) reinforced N-USP exhibited the lowest $K_s$ at all loads/abrading distances. It was observed that the wear rate of a solid body in three-body abrasion strongly depends on the hardness of the counter-body.

Figure 5.6a shows the specific wear rate of pure and hybrid composite under 24 N applied load. Pure coconut sheath polyester composite has the least specific wear rate. This is due to the good abrasive resistance of the coconut sheath fiber. In the beginning, the soft polymer medium was removed and then the hard coconut sheath fiber brought into contact. Hence, further wear was reduced. In this plot, NNG-USP composite showed higher wear rate in the beginning; this is due to the aggressive removal of the glass fibers and soft polymer materials in the beginning. After the whole erosion of the glass fibers, the hard coconut sheath fibers came into contact with the abrasive particle and rubber wheel and hence the wear was controlled.

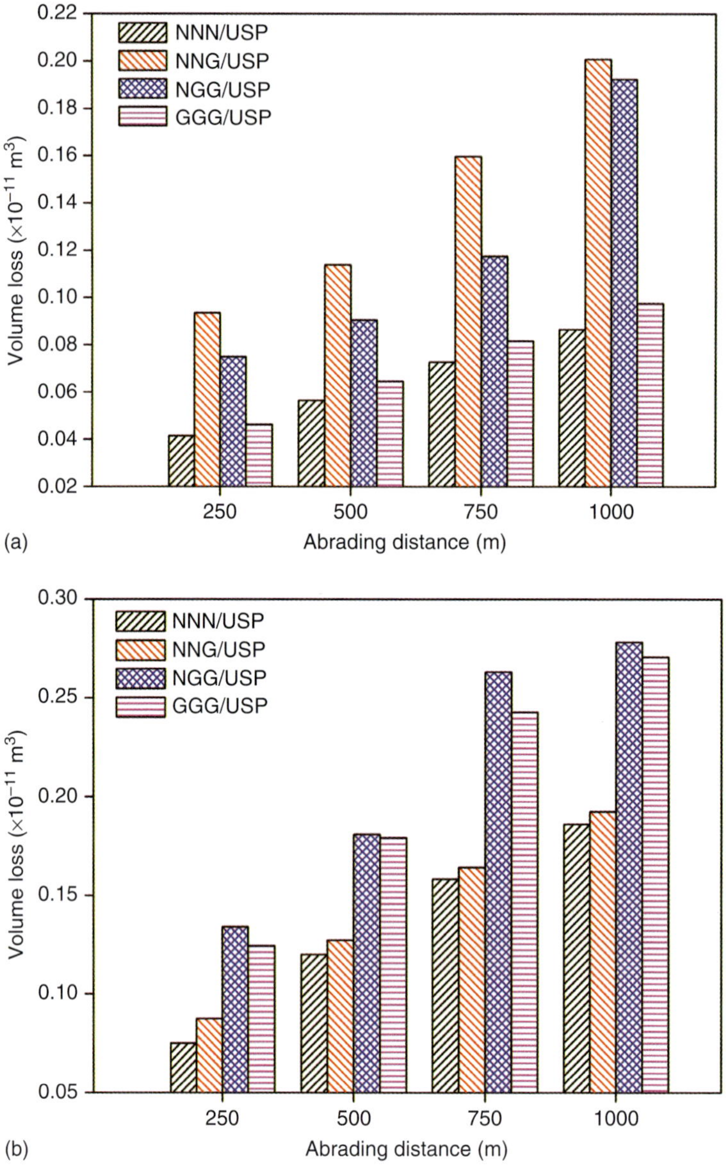


**Figure 5.5** Volume loss of pure and hybrid composites, run at 24 N (a) and at 36 N (b).

Similarly, in Figure 5.6b, NNN-USP composite showed the least specific wear among all composites. At higher applied load, the wear of the glass fiber composite was huge.

This is because abrasive particles can be embedded into the softer surface and then form grooves in the harder one. Severe damage to the glass fiber and USP matrix is the main reason for the higher wear rate of the GGG-USP composite. More debris and micro-cracks in the matrix were evident due to deformation of the softened matrix under higher load conditions. In this case, fiber

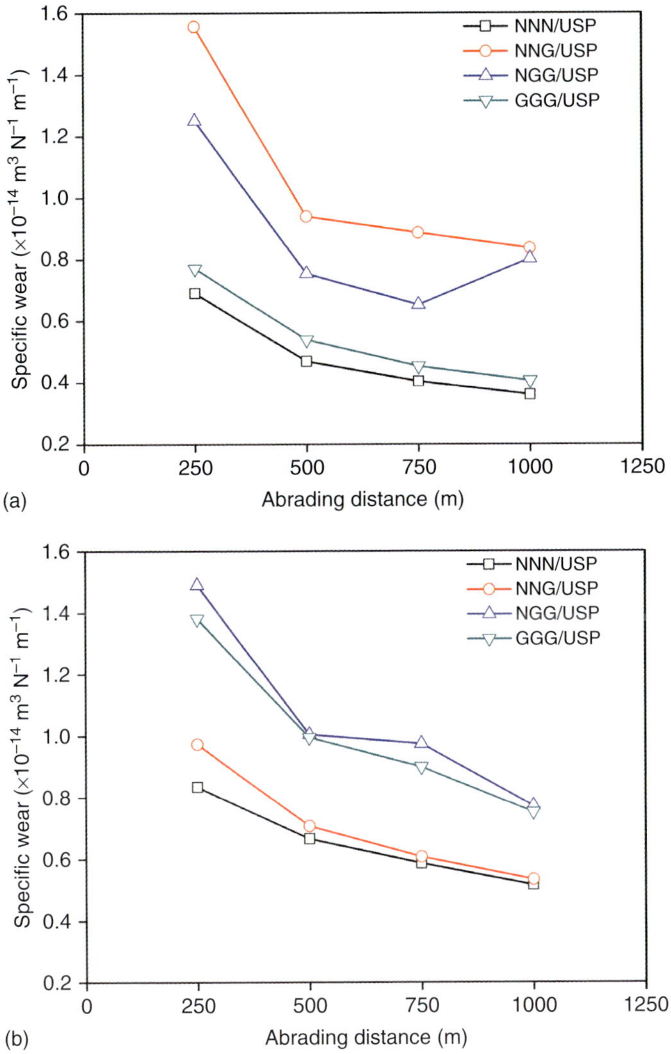


**Figure 5.6** Specific wear of pure and hybrid composites, run at 24 N (a) and at 36 N (b).

removal was aggravated since the softened polymer matrix could not effectively protect the brittle nature of glass fiber from peeling off. As a result, the wear resistance of the GGG-USP decreased significantly. From Figures 5.5 and 5.6, it is clear that the hybrid reinforced composites (GGN-USP and NNG-USP) showed higher specific wear rate, and the effect of hybridization is not beneficial from the abrasion resistance point of view. The specific wear rate increased when hybrid fibers were used in USP. As such, pitting cracks and fracture on the composite are evidenced, which lead to high material removal (Figures 5.3–5.6).

Figure 5.7 shows the worn surfaces of the pure and hybrid composites under 24 N applied load. Figure 5.7a reveals the debonding (DB) in the NNG-USP

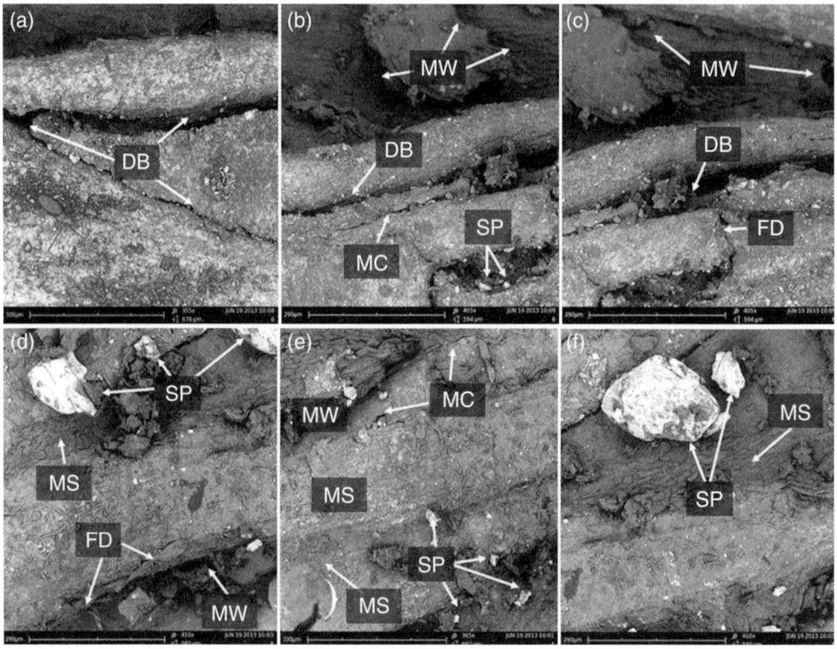


**Figure 5.7** (a–f) SEM images of worn surface of N-USP after wear (for 36 N of load and abrading distances of 250, 750, and 1000 m).

composites. In Figure 5.7b, wear on matrix (MW) formed due to matrix distortion and damage. The abrasive particles led to micro-ploughing (MP) on the fiber–matrix interface, and the cutting action due to the sharp tip of the abrasive particles (SP) damaged the fibers (FD) (Figure 5.7c). Hence, in most places the sharp abrasive particles plug into the fiber/matrix surfaces (Figure 5.7d). Such an observation reveals the matrix shrinkage (MS) and fiber damage with the matrix and hence the locking of smaller abrasive particle into the fiber–matrix interface occurred and micro-cracking (MC) was also observed (Figure 5.7e). Owing to the abrasive action, the fiber surface also deteriorated with matrix shrinkage (Figure 5.7f).

Figure 5.8 shows the worn surface of the pure and hybrid composites under 36 N applied load. From Figure 5.8a,b silica particle (SP) and micro-cracking (MC) damaged the surface due to the abrasive flow with striking pressure. At higher applied load also, cavities occurred on the fiber–matrix interface due to the micro-ploughing (MP) of the matrix materials by sharp abrasive particles (Figure 5.8c). Owing to the cutting action of the short abrasive particle, a progressive growth of cracks (CI) was noted on the fiber–matrix interface and the fiber damaged (FD) (Figure 5.8d). Lower matrix ductility eventually resulted in higher wear rate. Owing to higher applied load, the abrasive particles fed over tends to fill the micro-cracks which resulting micro-ploughing can be seen on the worn surface (Figure 5.8e,f).

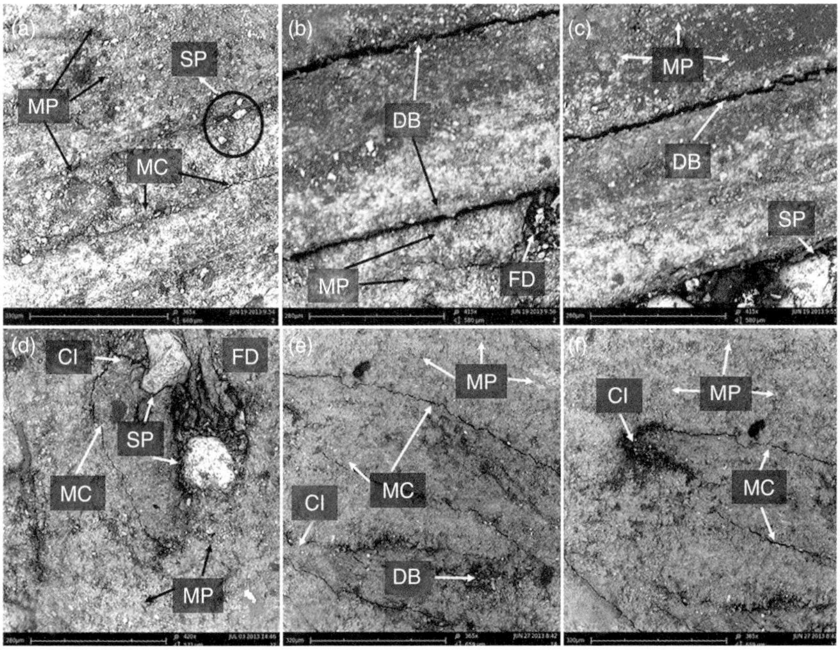


**Figure 5.8** (a–f) SEM images of worn surface of NNG-USP after wear (for 36 N of load and abrading distances of 250, 750, and 1000 m).

## 5.4 Conclusion

The mechanical and tribological properties of the coconut sheath and glass fiber reinforced USP matrix composites have been investigated. The effect of surface-modified fibers using a silane coupling agent on the tribological behavior is discussed. The following conclusions could be drawn:

- The coconut sheath and glass fiber reinforced USP matrix composites possess good tensile and hardness properties. There is also synergy effect between the coconut sheath and glass fibers on the enhancement of the wear resistance.
- Three-body abrasive wear of all the composites depends on the experimental parameters such as load, abrading distance, and type of fiber and also on loading of fibers in the USP resin.
- Glass fiber reinforced USP and GGN/NNG reinforced USP composites experienced higher specific wear rate under all test conditions employed in the present work.
- The best wear resistant composition achieved in this study is that of a silane treated coconut sheath fibers reinforced USP composite. Therefore, in abrasive wear situation, the use of synthetic glass fiber reinforcement or hybridization of fibers is not beneficial in the design of wear resistant polymer composites.

## Acknowledgement

The authors want to thank CNPq for the BJT grant and DST-CNPq for the support of the Brazil-India collaboration project.

## References


1 Scarponi, C. and Messano, M. (2015). Comparative evaluation between E-glass and hemp fiber composites application in rotorcraft interiors. *Composites Part B: Engineering* 69: 542–549.
2 Mansor, M.R., Sapuan, S.M., Zainudin, E.S. et al. (2013). Hybrid natural and glass fibers reinforced polymer composites material selection using analytical hierarchy process for automotive brake lever design. *Materials and Design* 51: 484–492.
3 Van Vuure, A.W., Baets, J., Wouters, K., and Hendrickx, K. (2015). Compressive properties of natural fibre composites. *Materials Letters* 149: 138–140.
4 Arthanarieswaran, V.P., Kumaravel, A., and Kathirselvam, M. (2014). Evaluation of mechanical properties of banana and sisal fiber reinforced epoxy composites: influence of glass fiber hybridization. *Materials and Design* 64: 194–202.
5 Jayamani, E., Hamdan, S., Rahman, M.R., and Bakri, M.K.B. (2014). Comparative study of dielectric properties of hybrid natural fiber composites. *Procedia Engineering* 97: 536–544.
6 Zhang, Y., Li, Y., Ma, H., and Yu, T. (2013). Tensile and interfacial properties of unidirectional flax/glass fiber reinforced hybrid composites. *Composites Science and Technology* 88: 172–177.
7 Winowlin Jappes, J.T., Siva, I., and Rajini, N. (2012). Fractography analysis of naturally woven coconut sheath reinforced polyester composite: a novel reinforcement. *Polymer-Plastics Technology and Engineering* 51 (4): 419–424.
8 Satyanarayana, K., Pillai, C., Sukumaran, K. et al. (1982). Structure property studies of fibres from various parts of the coconut tree. *Journal of Materials Science* 17: 2453–2462.
9 Siva, I., Winowlin Jappes, J.T., and Suresha, B. (2012). Investigation on mechanical and tribological behavior of naturally woven coconut sheath-reinforced polymer composites. *Polymer Composites* 33 (5): 723–732.
10 Siva, I. and Winowlin Jappes, J.T. (2011). Studies on the influence of silane treatment on mechanical properties of coconut sheath-reinforced polyester composite. *Polymer-Plastics Technology and Engineering* 50 (15): 1600–1605.
11 Harish, S., Peter Michael, D., Bensely, A. et al. (2009). Mechanical properties evaluation of natural fiber coir composite. *Journal of Material Characterization* 60 (1): 44–49.
12 Stachowiak, G.W. and Batchelor, A.W. (2000). *Engineering Tribology*. Amsterdam: Elsevier.
13 Shipway, P.H. and Ngao, N.K. (2003). Microscale abrasive wear of polymeric materials. *Wear* 255: 742–750.
14 Chand, N., Naik, A., and Neogi, S. (2000). Three-body abrasive wear of short glass fibre polyester composite. *Wear* 242: 38–46.

15 Irullappasamy, S., Durairaj, R., Irulappasamy, S., and Manoharan, T. (2016). Investigation on wear behaviors and worn surface morphology of surface treated palmyra fruit fiber/polyester composites to appraise the effects of fiber surface treatments. *Polymer Composite* doi: 10.1002/pc.24163.

16 Brahmakumar, M., Pavithran, C., and Pillai, R.M. (2005). Coconut fibre reinforced polyethylene composites: effect of natural waxy surface layer of the fibre on fibre/matrix interfacial bonding and strength of composites. *Composites Science and Technology* 65 (3–4): 563–569.

17 Azwa, Z.N., Yousif, B.F., Manalo, A.C., and Karunasena, W. A review on the degradability of polymeric composites based on natural fibres. *Materials and Design* 47: 424–442.

18 Yousif, B.F., Nirmal, U., and Wong, K.J. (2010). Three-body abrasion on wear and frictional performance of treated betelnut fibre reinforced epoxy (T-BFRE) composite. *Materials and Design* 31 (9): 4514–4521.

# 6

# Wear Properties of Acid and Silane Modified CNT Filled Hybrid Glass/Kenaf Epoxy Composites

*Napisah Sapiai, Aidah Jumahat, and Jamaluddin Mahmud*

*Universiti Teknologi MARA, Faculty of Mechanical Engineering, Shah Alam 40450, Selangor, Malaysia*

## 6.1 Introduction

Wear could be described as the surface behavior of a material when being subjected to continuous motion during interaction with another counterface. The quantity that relates closely to the wear of the material is the mass loss of the material after interaction. If the mass loss of a material is high when being grazed or abraded, it denotes that the material wears easily or the wear rate of the said material is high. There are various mechanisms of wear, such as the fiber–matrix debonding, the pulling out of fiber, the breakage of fiber and/or matrix, and the resin peel-off [1]. In polymer composites, the strength of the adhesion bond between the matrix and fiber could affect the adhesive wear of the composites. Higher adhesion of the bond will help in reducing the adhesive wear effect since higher bond strength suggests that the fiber will grip onto the matrix from being pulled out by the counterface [2, 3].

Abrasive wear is one type of wear where less defect of the material surface is observed when compared to a more defective wear test that could eventually end up breaking the matrix or fiber. There are four mechanisms of abrasive wear – micro-cutting, micro-ploughing, micro-fatigue, and micro-cracking – as shown in Figure 6.1. Micro-ploughing occurs when a material is displaced sideways by an abrasive particle, and the presence of ridges and grooves would be evident. Micro-cutting could be seen when there is an apparent detachment of particles when being swept. Micro-fatigue happens when the particles are removed after a repetitive encounter with the abrasive particles, and lastly, micro-cracking can be observed on the surface of the material. Micro-cracking usually can be observed on brittle materials such as ceramics and it is caused by the high stress concentration on the material surface, which is imposed by the abrasive counterface [4].

There are many factors that could affect the wear performance of a composite such as the working or test parameters, the structure of the composite, the materials used as matrix, and the type of reinforcement, namely, the material, the type of reinforcement (filler or fiber), the orientation of the fiber, and the filler

*Synthesis and Tribological Applications of Hybrid Materials*, First Edition.
Edited by Mohammad Jawaid, Rajini Nagarajan, Jacob Sukumaran, and Patrick De Baets.

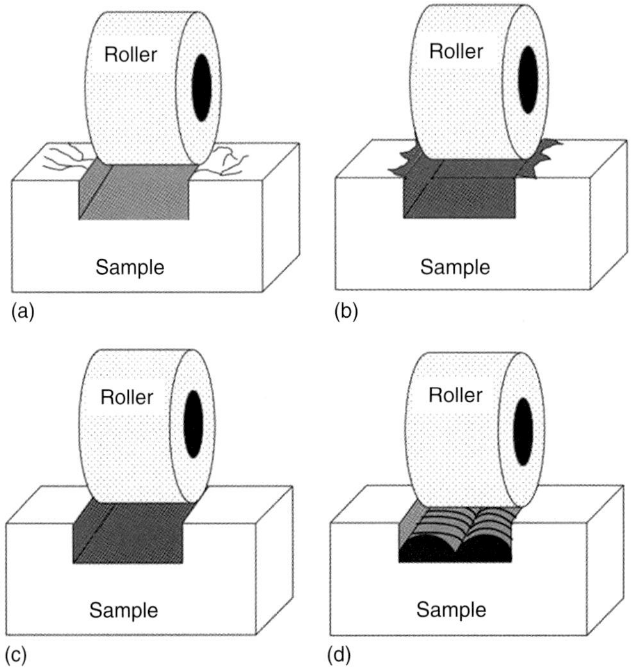


**Figure 6.1** Schematics of abrasive wear mechanisms: (a) micro-cracking, (b) micro-ploughing, (c) micro-cutting, and (d) micro-fatigue.

particles' size. The bond strength between the matrix and the reinforcement phase are also proved to be important factors in determining the wear properties. Earlier studies have shown that the test parameters indeed did affect the wear rates of the materials [3, 5–10]. Increasing test parameters such as the load, sliding speed, and distance will impose higher pressure contact on the materials' surface. This will generate more heat and friction and will induce wear. More heat generated could result in the softening of the matrix, which will then lead to easier peeling of the resin. Higher friction could cause a higher amount of material being removed at once [3, 6–8]. The working or testing environment also plays a crucial role in the wear rate. As was studied by Dhieb et al. [10], the environment could either benefit or worsen the wear properties of the material. In wet environment, the debris produced will be washed away so that the friction and heat generated would be lessened. However, for a material that produces smoother and smaller debris, this may not be beneficial at all. Usually, when a material produces smaller and smoother debris, a tribo-protective layer will form and it will act as the barrier between the materials' surface and the counter face. The barrier will help in reducing the friction and prevent more material from being swept away. Therefore, in this case, the environment might not help in improving the wear properties at all [10].

There are many potential materials that could be used in order to improve the wear properties of composites. Various studies have been carried out in order to discover the potential of natural fibers to be used as reinforcement in order to improve wear [3, 11, 12]. It was revealed that kenaf fiber has a good potential in reducing wear because it helps to reduce the adhesive wear of the matrix. Kenaf is also hard to be pulverized when in normal orientation to the counter face, thus making it favorable to be used in wear-affected applications [11, 12]. On the other hand, studies carried out on glass fiber show that glass fiber is easily crushed although it is known to be a good reinforcement in order to improve mechanical properties [2, 3, 8]. The pulverized glass fiber may induce wear as the debris produced may be involved in enhancing unwanted friction. The increase in friction will eventually result in worse wear rates [3, 8]. Therefore, in order to improve the wear behaviors of the glass fiber reinforced polymer composites, particles or fillers could be incorporated to help cushion asperities from shock and help prevent further wear of the material [3]. Many kinds of research were conducted in order to explore the possibilities of fillers in aiding to improve wear properties of composites [13–18]. It was found that nano-sized particles served better in improving wear resistance compared to micro-sized particles. Nanoparticles have a higher surface area that increases the area that could interact with the matrix, and hence, higher bond strength. Nanoparticles also help in producing small, smooth debris, which helps in forming the tribo-protective layer. This layer will help improve the wear resistance of the composite, specifically reducing the wear rates [13, 15, 16].

Over the years, many researchers have shown interest in the potential of carbon nanotubes (CNTs) as a wear resistance enhancer. However, it was highlighted by Yan et al. [18] that CNT tends to agglomerate easily if not being treated. The agglomeration will result in bigger chunks of CNT being swept away; thus the material loss will be higher, resulting in higher wear rate [18]. Therefore, treatments were done on the CNT in order to improve the dispersion and also the bonding with the matrix [19–21]. Acid treatment has proved to help in improving the dispersion of CNT and thus help boost the mechanical properties of the CNT. However, this treatment may have implications in the reduction of cross-linking network density and possibly interfere with the functional group of the CNT. On the other hand, silane modification will help in improving dispersion as well as in the formation of covalent bonds. Silane will also help improve thermal stability, flexural modulus, and strength of the composite; however, it will affect the reduction of electrical conductivity of the CNT [19, 21].

From these reviews, it could be found that no study has been reported pertaining to the effect of incorporating acid and silane treated CNT on the wear properties of epoxy/kenaf/glass polymer composites. Therefore, this paper aims to investigate the effect of incorporating acid and silane treated CNT on the wear properties of epoxy/kenaf/glass polymer composites.

## 6.2 Methodology

### 6.2.1 Materials

The matrix, Commercial epoxy (Miracsat 1517 A/B), was supplied by Miracon (M) Sdn Bhd, Malaysia. The yarn kenaf and glass fiber used were supplied by

Innovative Pultrusion Sdn Bhd, Seremban, Malaysia. The multi-wall carbon nanotubes (MWCNTs) (Flo Tub 9000 Series) used in this study were synthesized using the catalytic vapor deposition process. Its average diameter and length are 11 nm and 10 μm respectively. The CNT was supplied by CNano Technology (Beijing) Ltd. The reagents used in the acid treatment are sulfuric acid (95%, R&M Chemicals), nitric acid (62–65%, R&M Chemicals), acetone (99.5%, R&M Chemicals), and ethanol (99.5%, R&M Chemicals).

### 6.2.2 Fabrication of Epoxy/Kenaf/Glass Fiber/CNT Composites

#### 6.2.2.1 Acid Treatment and Silane Modification Process on CNT

Before being added to the epoxy resin, the CNT was first treated with acid and silane. Firstly, 3.0 g of unmodified CNT was weighed and dispersed in 300 ml of acid solution with a ratio of 3 : 1 ($H_2SO_4$:$HNO_3$). Next, the mixture was refluxed at 80 °C for 4 h using a hot-plate magnetic stirrer device. Next, distilled water was added to the solution and continuously stirred for another 6 h. The solution was then filtered with distilled water and acetone until a pH value of 6–7 was attained. The oxidized CNT was dried in an oven for another 24 h at 80 °C.

For silane modification, 1.8 g of the previously acid treated CNT powder was dispersed in 300 ml ethanol–water–silane solution. The solution was diluted with 2 wt% of *three-aminoprophyltriethoxysilane* to an aqueous solution of 300 ml ethanol–water (95 : 5). The mixture was stirred for 4 h at 70 °C. Then, the mixture was filtered and dried in the oven at 80 °C for 12 h.

#### 6.2.2.2 Fabrication of the Composites

Initially, a series of pristine carbon nanotubes (PCNTs) with several weight fractions of 0.5, 0.75, and 1.0 wt% were mixed with epoxy using the mechanical stirrer at 400 rpm for 1 h. Then, the mixtures were degassed in a high vacuum machine for another 1 h. Hardening agent with a ratio of 100 : 30 (epoxy:hardener) was added to the mixture. The degasification process was crucial in order to remove any trapped air that may have been produced unintentionally during the fabrication process. The whole process was then repeated using the same weight fraction series of acid treated carbon nanotubes (ACNTs) and silane treated carbon nanotubes (SCNTs).

The composites were produced using a combination of dry filament winding and impregnation of woven glass fiber and kenaf fiber into the epoxy, which had been incorporated with CNT beforehand. Kenaf fibers were wound on a frame of 430 × 300 mm in one direction. The woven glass fiber was layered on top of and underneath the wound kenaf fibers. After that, the hybrid glass/kenaf composites were left at room temperature for 24 h before being post cured in the oven at 60 °C for another 2 h, followed by 2 h at 80 °C, 100 °C for 2 h, and finally 120 °C for 2 h.

### 6.2.3 Abrasive Wear Test

The plates of composites were cut into disk specimens according to ASTM D3389. The abrasive wear test was conducted using the abrasive wear tester

(TR 600). A constant load of 10 N with 2.5 m $s^{-1}$ of sliding speed was applied. The sliding distance was set to 20 000 m with an interval of 2000 m [3].

For every 2000 m interval, the test was stopped and the mass of the specimen was weighed. In the interval, the roller was scrubbed and dusted in order to ensure a clean surface without any other foreign objects stuck onto it. At 10 000 and 20 000 m sliding distance, the thermal images were captured using an infrared thermal camera. With the mass loss recorded, the specific wear rate was then calculated using the following formula (Equation (6.1)):

$$W_s = \frac{\Delta m\,(\mathrm{g})}{L\,(\mathrm{m}) \times \rho\left(\mathrm{g\ mm^{-3}}\right) \times F\,(\mathrm{N})} \tag{6.1}$$

where the $W_S$ denotes the specific wear rate and was described as a function of mass loss ($\Delta m$) over the multiplications of sliding distance, $L$, density ($\rho$), and applied load ($F$). The specific wear rates are expressed in $mm^3\ N^{-1}\ m^{-1}$ [12].

### 6.2.4 Characterization of the Abraded Surface of the Composites

The abraded test specimens were then further analyzed using the stereomicroscope with 10× magnification. The surface profile of the abraded specimens was analyzed using Alicona imaging technique.

## 6.3 Results and Discussion

### 6.3.1 The Effect of Incorporating PCNT to Hybrid Glass/Kenaf Composites on the Wear Properties of the Composites

As described by Chauhan and Thakur [22], the steep reduction in mass loss during the initial part of the graph as seen in Figure 6.2 is due to the difference in the surface roughness of the specimens' surface and the counter face. They also stated that the difference in surface roughness allowed a strong interlocking process to occur, thus leading to larger amounts of material to be removed, and this explained why in the beginning the wear rate was high. For all types of composites, it could also be seen that after 15 000 m of sliding distance, the wear rate started to be in a steadier state; after being exposed to the abrasive counter face until the said distance, the surface topography became more and more similar to the counter face, which had weakened the interlocking process, thus resulting in lower wear rate. This behavior will be observed continuously until the minimum value of the wear rate is obtained [23].

Figure 6.2 also shows that when glass fiber was added into kenaf/epoxy matrix, the specific wear rates increased in a significant manner. When being subjected to the abrasive roller, after the epoxy layer was removed about 1 mm in depth, the glass fiber layer was exposed. Since it was reported by Ruckdäschel et al. [23] that glass fiber could easily be pulverized, it could be explained that the increase in material loss was due to the pulverization of glass fiber. From the thermal images captured during abrasion test it could be observed that the glass composite

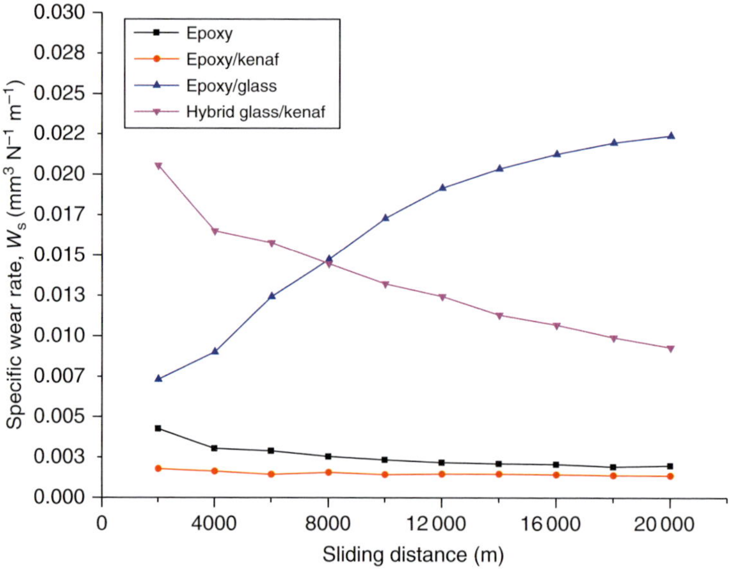


**Figure 6.2** Graph of specific wear rates of different types of samples.

produced higher contact temperature (Figure 6.3b) than epoxy polymer (Figure 6.3a). This supports the claim that the glass fibers were being pulverized vigorously, thus causing higher friction, which resulted in significant increment in temperature. It should be noted that the temperature did not increase immediately after being in contact with the abrasive roller; instead, the temperature was increasing cumulatively with respect to the increasing amount of rotations.

Figure 6.2 also shows that when glass was hybridized into kenaf composites, the wear rates increased compared to the kenaf composite. However, the wear rates were actually lower compared to glass fiber composites. Since the glass fibers were layered on top of the kenaf, the glass fiber wore out first. It was because the glass fiber was in the woven mat form; according to Ruckdäschel et al. [23], the surface topography of the glass fiber must have increased the intensity of the interlocking of asperities, which enabled larger amounts of glass

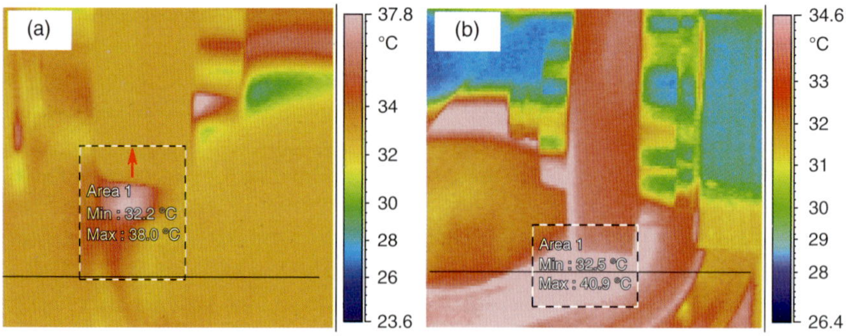


**Figure 6.3** Thermal images of (a) epoxy and (b) glass composites.

fiber to be pulverized. Any sign of decrease in wear rate for both the glass fiber composite and hybrid glass/kenaf composite could not be seen as the layer of glass fiber had not yet been fully removed. The kenaf layer had yet to be abraded.

Figure 6.4 illustrates the images obtained from stereomicroscope to observe the surface of the abraded specimens. The black arrows indicate the sliding directions of the sample. It can be seen based on Figure 6.4a that the abraded surface of the epoxy resin appeared to be rough. The rough abrasion may induce larger debris to form, thus inducing the three body abrasive wear [17]. The rough worn surface indicates that apart from the apparent micro-cutting and micro-ploughing, the sample was also being subjected to micro-fatigue abrasive wear, which is also one of the basic mechanisms of abrasive wear [4, 18].

The abraded surface of glass composite as seen in Figure 6.4c appeared to be rougher compared to epoxy. Since glass fibers were known to be easily pulverized, the crushed bits of the glass fibers alongside the debris of epoxy may have increased the effect of three body abrasive wear. This will lead to higher wear rate and friction, thus agreeing with the current findings. The difference in surface topography provided by the woven glass fiber contributes to increase the wear

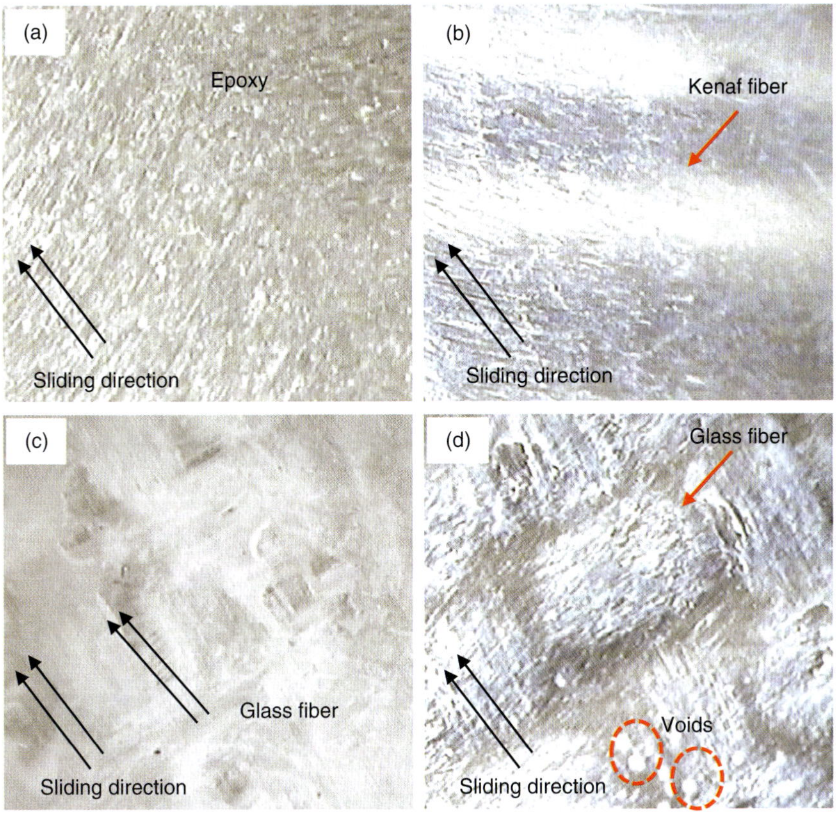


**Figure 6.4** Images from stereomicroscope for (a) epoxy, (b) kenaf composite, (c) glass composite, and (d) hybrid glass/kenaf composites, magnification 10×.

rate, as the difference in surface topography provides better interlocking between the surface and the counter face, which will lead to easier digging and ploughing of the surface [22, 23]. The worn surface for hybrid glass/kenaf composites appeared to be smoother compared to glass composites as seen in Figure 6.4d. The presence of kenaf fiber may have improved the adhesive wear effect of the composite, thus reducing the wear rate when compared to glass composites. After a sliding distance of 20 000 m, it can be observed that the remaining glass fiber still intact in the composites. This explains why the wear rates of epoxy/kenaf/glass composite could not be considered as an improvement compared to epoxy resin although it was better than epoxy/glass composite.

Figure 6.5 displays the surface condition of the abraded surfaces of kenaf composite, glass composite, and hybrid glass/kenaf composites. Similar to Figure 6.4, the surface roughness of kenaf composite was observed to be the smoothest followed by hybrid glass/kenaf composite and glass composite. In this case, voids could be detected as highlighted in Figure 6.5. The existence of voids contributed in the increase of the surface topography, which will further induce wear [23]. The voids also represent inconsistency, which may vary the results obtained. The voids appeared larger in the glass and hybrid glass/kenaf composites. This may

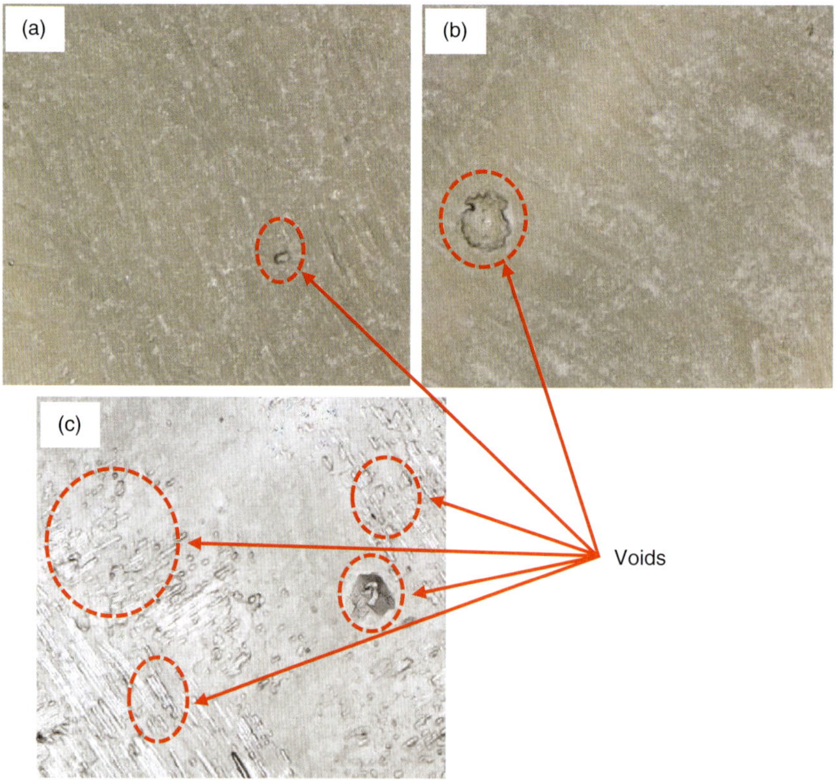


**Figure 6.5** Surface images of (a) kenaf composite, (b) glass composite, and (c) hybrid glass/kenaf composites obtained from Alicona 3D surface metrology.

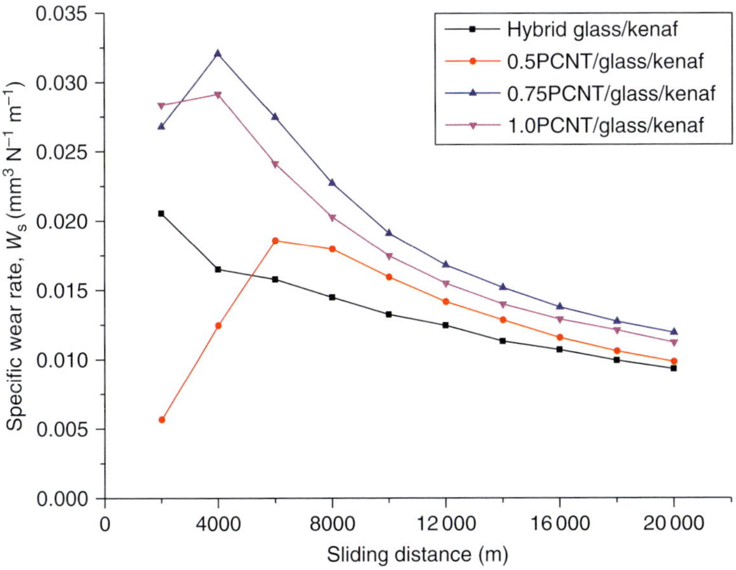


**Figure 6.6** Effect of hybridizing PCNT into hybrid glass/kenaf composite on the wear rate.

be due to the difficulties for the epoxy resin to properly wet the woven glass and kenaf fibers due to its viscosity and also the structure of the glass fiber, hence inducing the formation of voids.

When CNT was added into hybrid glass/kenaf composites, the wear rates were not improved compared to the hybrid glass/kenaf composite as shown in Figure 6.6. Instead, when added with CNT, the wear rates increases. The highest increment was observed for 0.75 PCNT/glass/kenaf, followed by 1.0 PCNT/glass/kenaf and 0.5 PCNT/glass/kenaf. There are many reasons that could contribute to these findings. One of them would be that the CNT had not been dispersed well. Since glass fiber was also used, this reduced the ratio of epoxy in order to accommodate the inclusion of CNT and glass fiber. This may lead to improper wetting of the fibers as well as fillers. Higher exposure of the fibers and fillers caused by improper wetting may lead to easier digging and ploughing action, which resulted in higher wear rate [24].

Thermal images of the PCNT filled glass/kenaf composites are shown in Figure 6.7. The contact temperatures for the PCNT filled glass/kenaf composites were recorded to be (a) 35 °C for 0.5 PCNT/glass/kenaf, (b) 34.0 °C for 0.75 PCNT/glass/kenaf, (c) 32.8 °C 1.0 PCNT/glass/kenaf, and (d) 35.0 °C for hybrid glass/kenaf composites. When compared to glass composite, the contact temperature of PCNT filled glass/kenaf composites dropped significantly. The contact temperature of glass/kenaf composite had dropped significantly when compared to glass composite. The addition of kenaf may have improved the thermal stability and improved the adhesion wear resistance, thus producing smaller sized debris during abrasion. The smaller sized debris produced had reduced the friction caused during abrasion, thus lowering the contact temperature [3, 11, 25]. On the other hand, the inclusion of PCNT into the glass/kenaf composites

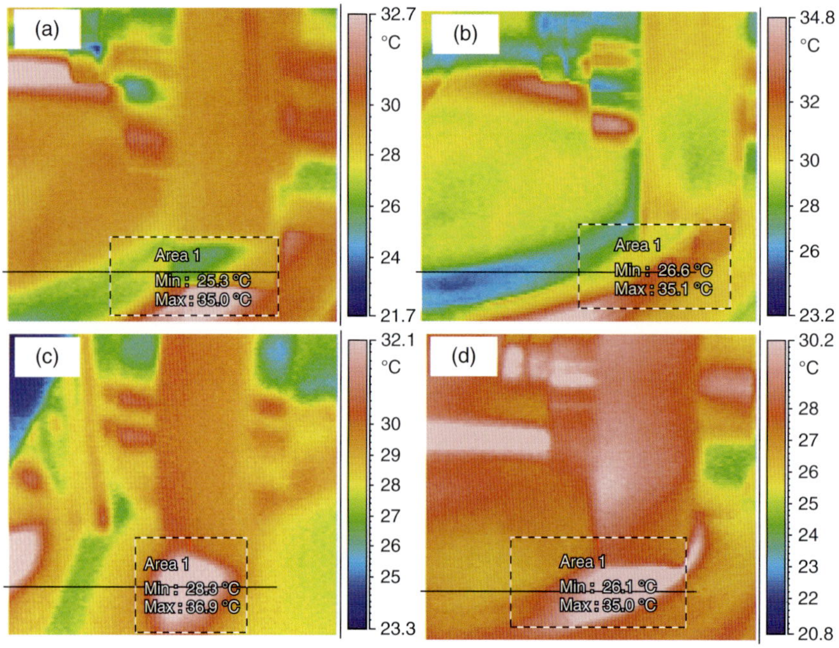


**Figure 6.7** Thermal images of (a) 0.5 PCNT/glass/kenaf, (b) 0.75 PCNT/glass/kenaf, (c) 1.0 PCNT/glass/kenaf, and (d) hybrid glass/kenaf composites.

showed small decreases in the contact temperature compared to glass/kenaf composites. The PCNT may have generated a tribo-protective layer, which had reduced the friction caused by the pulverization of glass fibers. The amount of PCNT included was very small, making the changes observed not very substantial [18].

The surface profiles of the abraded surface for the PCNT filled glass/kenaf composites are shown in Figure 6.8. The abraded surface for 0.5 PCNT/glass/kenaf is smoother compared to 1.0 PCNT/glass/kenaf. Although 0.75 PCNT/glass/kenaf has the smoothest abraded surfaces, many voids could be seen on the surface of the composite. These voids may have induced the wear rates, as it was discussed before that the different surface topographies will enhance the interlocking process, thus resulting in higher wear rates [23]. It could also be highlighted that there were some unscratched spots that may have been caused by the uneven surfaces or the misalignment of the abrasive rollers. These unscratched spots may have resulted in inconsistent results.

The surface profile was further analyzed using the Alicona technique and shown in Figure 6.9. Based on Figure 6.9, it could be observed that 0.5 PCNT/glass/kenaf has the smoothest surface profile compared to 0.75 PCNT/glass/kenaf and 1.0 PCNT/glass/kenaf. For 0.75 PCNT/glass/kenaf, there are many voids and scratches shown in the surface profile. The scratches may have been caused by the debris formed by the glass fiber, epoxy, and PCNT, which resulted in three body abrasive wear. The voids provide better interlocking, which may induce larger debris to form, leading to higher wear rates. The 1.0 PCNT/glass/kenaf surface

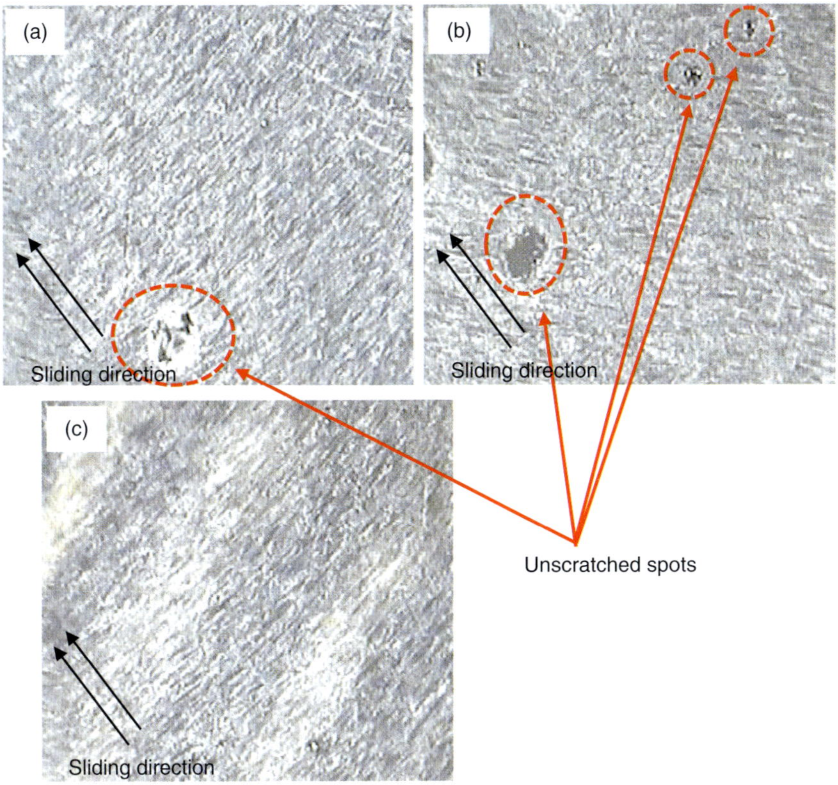


**Figure 6.8** Surface profiles for worn surfaces of (a) 0.5 PCNT/glass/kenaf, (b) 0.75 PCNT/glass/kenaf, and (c) 1.0 PCNT/glass/kenaf obtained from stereomicroscope.

was observed to have the roughest surface profile than the other two. Since 1.0 wt% was the highest amount of PCNT included into the hybrid glass/kenaf/composite, less epoxy was used. Since the epoxy used was the least, there was a higher chance of improper wetting of the fibers to occur, hence reducing the wear resistance of the hybrid glass/kenaf composites.

### 6.3.2 The Effect of Incorporating ACNT and SCNT to Glass/Kenaf Composite on the Wear Properties of the Composites

Next, the effects of acid treatment on the CNT for the ACNT filled hybrid glass/kenaf composites are shown in Figure 6.10. The acid treatment did not have any effect in improving the wear resistance of the CNT filled glass/kenaf composites. In fact, the wear rates increased when the CNT was treated with acid. The increment in wear rates may be due to the reduction of cross-linking density, which is one of the side effects of acid treatment. The decrease in bond strength will induce the adhesive wear, thus increasing wear rates [14, 19].

Figure 6.11 shows the effect of silane treatment on the CNT for SCNT filled glass/kenaf composites. The 0.5, 0.75, and 1.0 PCNT/glass/kenaf has lower wear rates compared to 0.5, 0.75, and 1.0 SCNT/glass/kenaf respectively. Although it

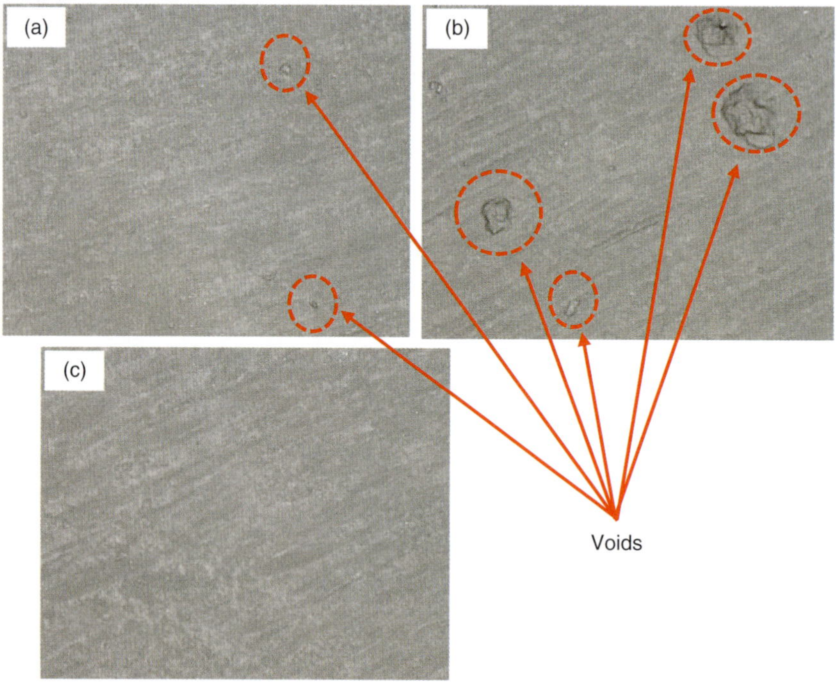


**Figure 6.9** Surface profiles for worn surfaces of (a) 0.5 PCNT/glass/kenaf, (b) 0.75 PCNT/glass/kenaf, and (c) 1.0 PCNT/glass/kenaf obtained from Alicona 3D surface metrology.

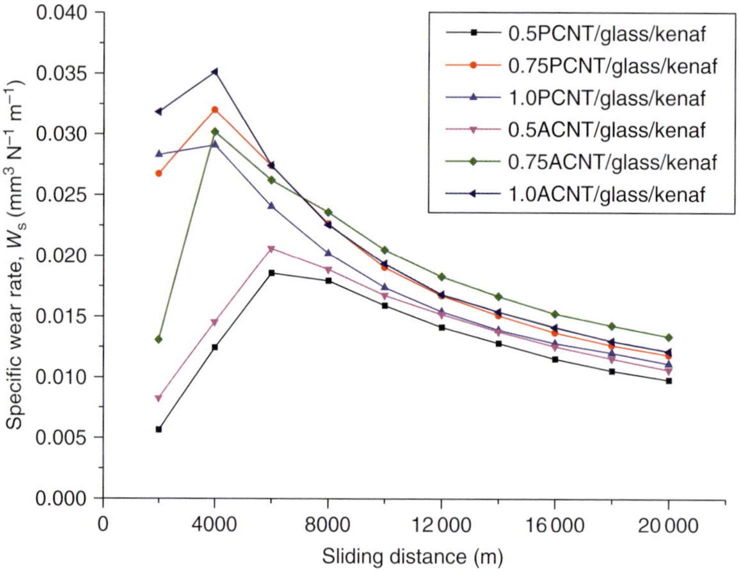


**Figure 6.10** Effect of acid treatment on the ACNT filled glass/kenaf composites.

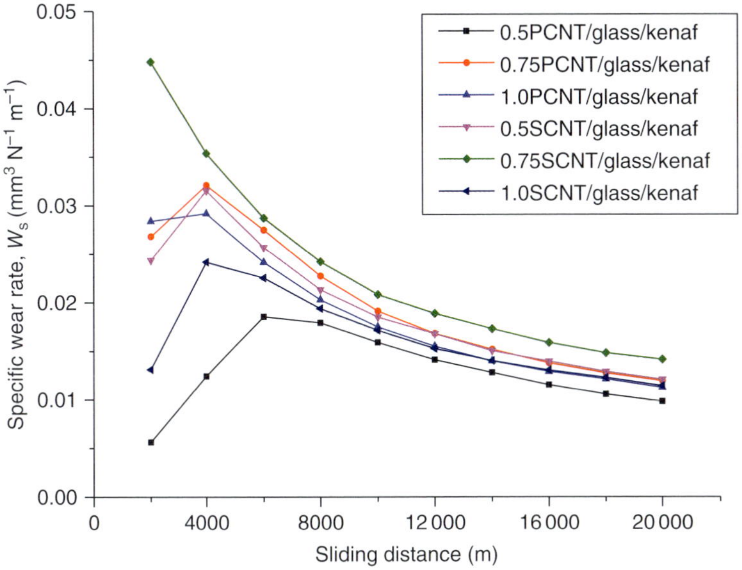


**Figure 6.11** Effect of silane treatment on the SCNT filled glass/kenaf composites.

has already been established in previous studies that silane treatment had improved the mechanical properties of the composites, it may have a conflicting effect on the wear properties. It was postulated that since silane had improved the interfacial bonding between CNT and epoxy, strong adhesion between CNT and epoxy may have caused large-sized and rougher debris to be ploughed. The production of large-sized and rougher debris will eventually lead to the increase of three body abrasive wear effect. This will induce the increase in wear rates [17, 20, 23].

Figure 6.12 shows the thermal images of the ACNT and SCNT filled glass/kenaf composites. As observed in Figure 6.12, the SCNT filled glass/kenaf composites recorded higher contact temperatures compared to their respective ACNT filled glass/kenaf composites. The temperature was also higher compared to PCNT filled glass/kenaf composites and glass/kenaf composite as recorded before in Figure 6.7. It may be deduced that the acid treatment may have disturbed the functional group, which led to the weakening of bonding strength between PCNT filler and epoxy matrix. The bonding strength was further reduced when the CNT was treated with silane since the covalent bonds have not been able to form due to the interrupted functional groups [19–21]. Since the bonding strength correlates to the wear rates, weaker bond strength as well as strong adhesion between CNT and epoxy contributes to higher wear rates. Higher wear rates may have produced larger debris, which will induce friction, resulting in increasing temperature. The weakened bond strength may also have induced adhesion wear, which leads to the increase of materials being removed, that is, higher mass loss and wear rates [1, 2, 14].

The surface profile for the ACNT and SCNT filled glass/kenaf composites are shown in Figure 6.13. The 0.5, 0.75, and 1.0 ACNT/glass/kenaf had similar

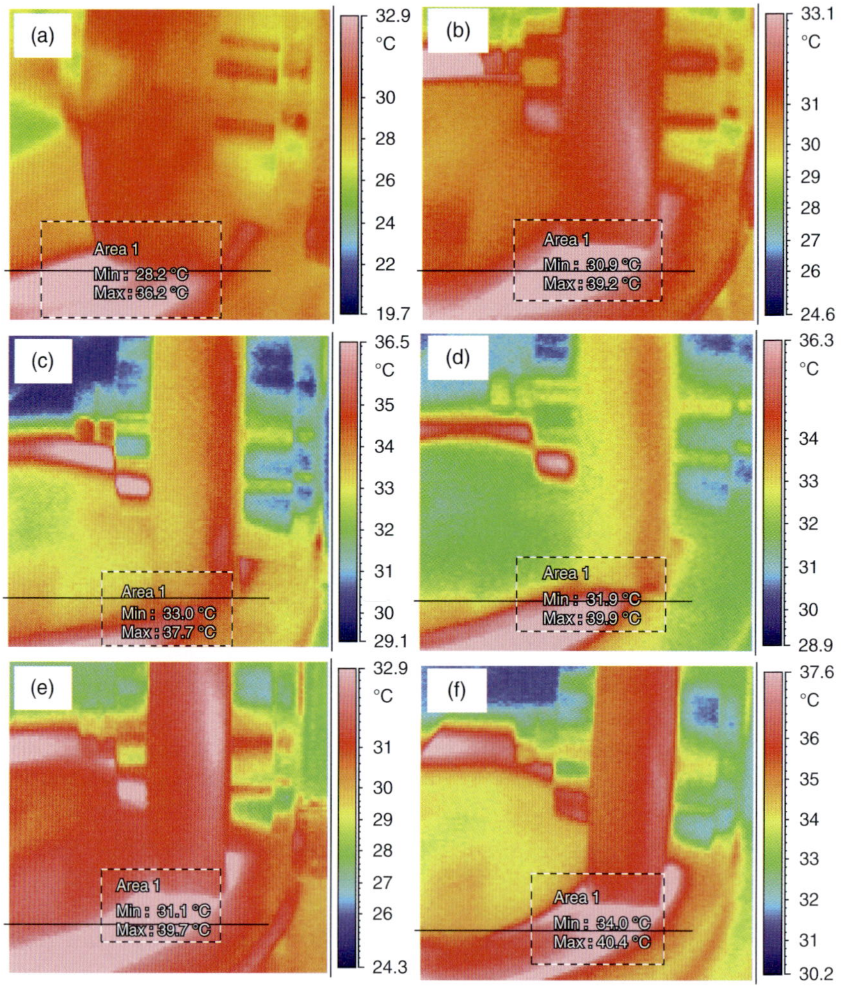


**Figure 6.12** Thermal images captured for (a) 0.5 ACNT/glass/kenaf, (b) 0.5 SCNT/glass/kenaf, (c) 0.75 ACNT/glass/kenaf, (d) 0.75 SCNT/glass/kenaf, (e) 1.0 ACNT/glass/kenaf, and (f) 1.0 SCNT/glass/kenaf.

surface profile compared to 0.5, 0.75, and 1.0 SCNT/glass/kenaf respectively. However, when compared to glass/kenaf composites surface profile from Figure 6.4d and PCNT filled glass/kenaf composites from Figure 6.8, the ACNT and SCNT filled glass/kenaf composites had rougher surface. Again, the rougher worn surface of the treated CNT composites may be due to the three body abrasive wear effect and micro-fatigue wear effect. These two effects are the results of the weakened bond strength between the acid-treated CNT and the epoxy, which may have induced the adhesive wear mechanism. Furthermore, the strong adhesion between silane-treated CNT and epoxy also may have resulted in the production of large-sized and rougher debris [14, 19, 21]. The reduction in bond strength induced the adhesion wear, where more material was being removed and the debris produced were bigger in size and

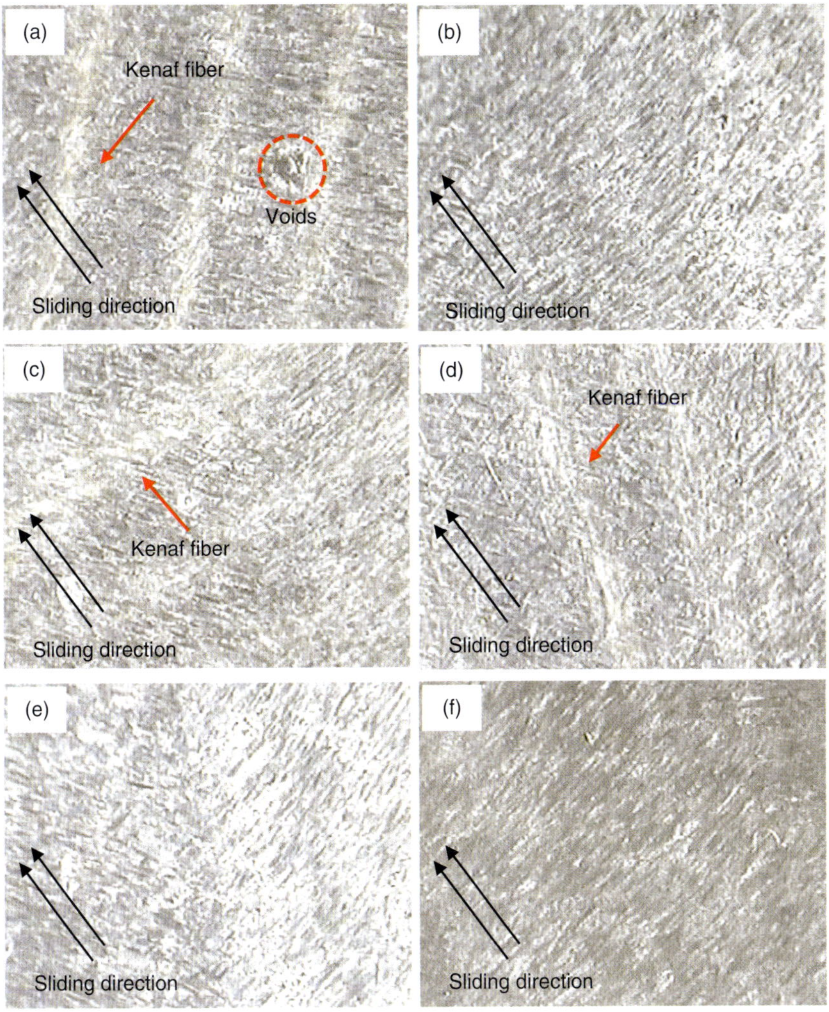


**Figure 6.13** Surface profiles for (a) 0.5 ACNT/glass/kenaf, (b) 0.5 SCNT/glass/kenaf, (c) 0.75 ACNT/glass/kenaf, (d) 0.75 SCNT/glass/kenaf, (e) 1.0 ACNT/glass/kenaf, and (f) 1.0 SCNT/glass/kenaf.

rougher in shape, thus causing the three body abrasive and micro-fatigue wear [1, 2, 14]. The Alicona images obtained for the ACNT and SCNT filled glass/kenaf composite showed no significant changes. The roughness of the surfaces also appeared similar to the PCNT filled glass/kenaf ones.

## 6.4 Conclusion

In conclusion, the main objectives of this study have been achieved successfully. The effect of incorporating PCNT, ACNT, and SCNT to hybrid glass/kenaf on the wear properties of the composites was studied. Since glass fiber was easily

**22** Chauhan, S.R. and Thakur, S. (2013). Effects of particle size, particle loading and sliding distance on the friction and wear properties of cenosphere particulate filled vinylester composites. *Materials and Design* 51: 398, 408.

**23** Ruckdäschel, H., Sandler, J.K.W., and Altstädt, V. (2018). On the friction and wear of carbon nano-fiber reinforced PEEK-based polymer composites. *Tribology and Interface Engineering Series* 55: 149–208.

**24** Sharma, S., Bijwe, J., Panier, S., and Sharma, M. (2015). Abrasive wear performance of SiC-UHMWPE nano-composites – influence of amount and size. *Wear* 332: 863–871.

**25** Guo, Q.B., Lau, K.T., Rong, M.Z., and Zhang, M.Q. (2010). Optimization of tribological and mechanical properties of epoxy through hybrid filling. *Wear* 269: 13–20.

# 7

# Hybrid Natural Fiber Composites as a Friction Material

*Patric D. Neis and Ney F. Ferreira*

*Federal University of Rio Grande do Sul, Department of Mechanical Engineering, Laboratory of Tribology, Av. Sarmento Leite, n. 425, Porto Alegre 90050-170, Brazil*

## 7.1 Friction Material Components

More than 2000 different ingredients are currently used in a brake friction material (brake pad or lining) [1]. Brake pads consist of a composite material, which is composed of a number of different compounds. Up to 20 or 25 different ingredients are usually added in a brake friction material [2–5]. According to the literature [3, 6], typical components used in brake friction materials include the following:

- *Structural materials (reinforcement)*: Reinforcement components provide mechanical strength. Usually metal fibers, carbon, glass, and/or aramid are used. More rarely, different mineral, ceramic, or natural fibers (coconut, sisal, and others) are employed. Asbestos was the most commonly used reinforcement component in the past. Nowadays, most of the countries have banned the use of asbestos.
- *Frictional modifiers:* These components are added to stabilize the friction force and to control the wear rates of both pad and disk. Solid lubricants (graphite, metal sulfides, etc.) are used to ensure stable frictional properties, especially at elevated temperatures. Abrasive particles, such as alumina and silica, increase both the coefficient of friction (COF) and the disk wear. Examples of friction modifiers include brass, zinc, copper, and graphite.
- *Binder*: They serve to hold the other components together and form a thermally stable matrix. Phenolic resins are commonly used as a binder, often with the addition of rubber for improving damping properties.
- *Fillers*: These components are usually used to reduce cost. Different minerals such as mica and vermiculite are often employed. Barium sulfate is another commonly used filler component.

*Synthesis and Tribological Applications of Hybrid Materials*, First Edition.
Edited by Mohammad Jawaid, Rajini Nagarajan, Jacob Sukumaran, and Patrick De Baets.

### 7.1.1 Friction Materials Requirements

As part of a commercial vehicle, brake friction materials (pad or lining) have numerous requirements, such as resistance to corrosion, light weight, long life, low noise and vibration, stable frictional property, low wear rate, no damage to the brake disk, and acceptable cost vs. performance [1, 7]. Besides, it is desirable to have a good friction performance in a wide range of temperature, contact pressure, and velocity [6]. Ostermeyer and Müller [8] added that a reduced amount of airborne dust is expected to be released from a friction material. Besides, due to recent restrictions on use of hazardous materials (e.g. U.S. regulations SB 6557 and SB 346), it is believed that a brake friction material should be also environmentally friendly.

Mustafa et al. [9] give numerous desirable characteristics for brake friction materials. As reported by the researchers, automotive brake friction materials should be strong enough to absorb brake torque even at high temperatures and various environmental conditions. They also added that high resistance is a desirable characteristic for prolonging periods of maintenance.

In friction materials, particularly in brake pads, heat is a critical factor. During a regular braking operation, the temperature on the rotor reaches up to 370 °C for the disk-pad of passenger cars [9]. In case of the standard procedure SAE J2522, the maximum temperature reached in a passenger brake car is even higher than the value reported by Mustafa et al. [9]. As described in the SAE standard, the maximum temperature of a disk brake system can reach up to 600 °C during the fade section of the referred standard procedure.

So, it is easy to understand how difficult it is to find an optimal formulation for friction materials, i.e. a formulation that is able to meet all the requirements described above. Besides, a series of laws and regulations are being adopted worldwide to promote the reduction of toxic and polluting elements from composites. Such elements can often be found as structural reinforcing fibers and/or additives in composite materials. A typical example of pollutant element is copper, which is commonly added to composite materials used in brake linings and pads, as well as in vehicle clutch linings. North American regulations, such as SB 6557 (brake friction material restriction on use) and SB 346 (hazardous materials: motor vehicle friction materials), have recently imposed restrictions on the addition of copper in friction materials composites. According to Lee and Filip [10], studies conducted by the US government have shown that copper is toxic to aquatic organisms. This is because exposure of silverfish (*Oncorhynchus kisutch*) to dissolved copper leads to a decrease in olfactory sensitivity of the species. This sensorial loss causes a reduction in the ability of this fish to avoid predators [11, 12], which may contribute to the extinction of this species. In this context, it is important to develop "clean" alternatives to substitute pollutants (such as copper) that are often added to composite materials.

Natural fibers are among the "clean" alternatives to substitute toxic and pollutant compounds in brake friction materials. However, information on the tribological behavior of composites made of natural fibers (sisal, bamboo, banana peels, coconut, etc.) is limited in the literature. The use of natural materials has enormous potential for application, since they are considered eco-friendly and therefore they are sustainable materials.

### 7.1.2 Braking Test Procedures

Until the middle of the 1950s, tests involving brake systems were only performed on vehicles in a type of test called "on road." However, this procedure is considered costly and time consuming due to the time required in setup and execution. Moreover, different drivers, changes in the environmental conditions, and traffic can make the analysis of the tests a hard task [13–15]. Nowadays, most tests involving brake friction materials are performed in laboratories. Tests performed on vehicles (on road tests) remain important, but they are limited to assessing the ergonomics and acoustic comfort of the brake system, as well as for confirming the results obtained in laboratory tests carried out with a brake dynamometer [16].

There are several kinds of machines designed for the purpose of testing friction materials from car brakes. As mentioned in some publications [14, 17] some testing machines are designed for quality control, while others are for performing basic research. Each testing machine used to assess the performance of friction materials has its own standard testing procedure. For instance, the test procedure for FAST machine (*Friction Assessment and Screening Test*) is described by SAE *J661*. In the case of a brake dynamometer (Figure 7.1), which is one of the most reliable test rigs used to test brake friction materials, there are many international standardized methods currently available. According to Papinniemi et al. [19], this machine can accurately represent the braking performance of a vehicle. Examples of standardized procedures designed for inertia dynamometers include SAE J2521, SAE J2522, FMVSS 135, and ISO 26867.

On the other hand, tribometers have stood out as a good alternative to perform basic research about friction and wear mechanisms of friction materials [20–23]. Basic research is important for the development of new friction materials, since the knowledge about tribology in brakes is still in initial stages compared to the evolution in other areas of automotive engineering. Besides, even today many brake manufacturers have been using empirical methods (trial and error) to develop new friction materials, as reported in some publications [4, 22].

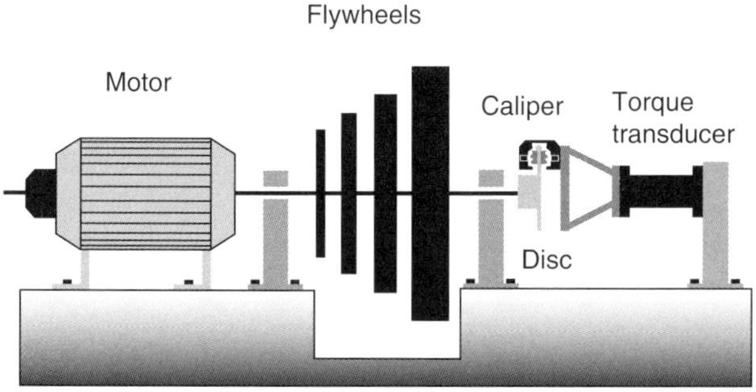


**Figure 7.1** Scheme of the brake dynamometer. *Source:* Haynes Publishing 1999 [18]. Reproduced with permission from Haynes Publishing.

Most standard procedures for braking tests have in common the sections described below:

*Green section*: This is the first stage of braking tests. It consists of performing a number of braking stops under constant parameters (normal load, sliding velocity, and initial temperature). This section aims to assess the brake performance of a friction material in a "green" condition. This is necessary because brake pads and linings are not completely cured during their development process.

*Running-in or bedding-in section*: This is characterized by a transitory raise in friction coefficient when the brake material wears out locally to conform to the surface of the disk [24, 25]. As explained by Kumar and Bijwe [24], running-in is an effect caused by the increase in the contact area. So, by means of a runing-in, it is possible to evaluate how long this transitory process lasts. A short running-in period is expected from a good brake pad.

*Pressure and velocity sections*: In these sections, braking stops are carried out under different levels of pressure and sliding velocity. In this way, the effect of these two parameters on wear and friction performance is assessed.

*Fade section*: According to Limpert [26], at high temperatures (>300 °C) a reduction in COF may occur. This effect is known as fading, or simply fade [26, 27]. As reported in [27], the causes of fading in brakes are usually attributed to the increase in the build-up of friction film. The degradation in the phenolic resin of the organic matrix composite is also related to the fade effect [21]. An excessive fading is undesirable [8] and may put the driver in danger. This is particularly important in extreme conditions, such as driving downhill for a long time [28].

*Recovery section*: This is the transient period that occurs immediately after braking stops at high temperature. During this section, the COF is expected to recover (and hence the name of the section), reaching its nominal magnitude. According to the literature [21], the reasons for the recovery include additional curing material, reversible and irreversible chemical changes, as well as vaporization of volatile compounds of the brake friction material.

## 7.2 Natural Fibers Used in Friction Materials Composites

Owing to the recent environmental laws and restrictions, friction material manufacturers show increasing interest in the use of natural fibers in brake pad and linings. Examples of such restrictions are the new regulations SB 6557 and SB 346 imposed by US regulators against adding copper into friction material composites.

Most natural fibers are nontoxic for humans and are also considered environmentally friendly. Examples of these natural materials include fibers of cotton, sisal, coconut, banana peels, palm kernel, etc. On the other hand, mineral fibers are considered hazardous materials, although they also come from nature. It is because they can affect human health during extraction. For this reason, mineral fibers are usually added in small amounts in composites [29]. Asbestos is the most typical example of a mineral fiber that is associated with numerous health problems. This material has been widely used in brake and clutch composites in the past. Reasons for this include its good heat and corrosion resistance, good mechanical properties, and low cost. Other common uses of asbestos are in steam pipes, gloves, ceiling tile, boilers, and furnace ducts insulated with an asbestos blanket or asbestos paper tape. Asbestos is a mineral fiber that can be easily identified only with optical microscope [30], as shown in Figure 7.2.

Six distinct types of asbestos have been identified in the literature [31]: amosite, actinolite, crocidolite, anthophyllite, tremolite, and chrysotile (Figure 7.3). Chrysotile represents around 93% of commercial use. All types of asbestos contain long chains of silicon and oxygen. This explains the fibrous nature of the mineral.

As mentioned above, asbestos is dangerous to human health. The asbestos fibers can penetrate the tissue, and cause serious health problems to humans, especially in the lungs. Lung cancer, pleural thickening, and asbestosis are among the most common sicknesses caused after heavy exposure to asbestos over many years. According to the World Health Organization (WHO) there is no safe asbestos. Also, according to the WHO, the best way to stop asbestos-related diseases is by banning its use worldwide.

According to information reported by the International Ban Asbestos Secretariat [32], there are currently 55 countries that have banned the use and commerce of asbestos. However, notable exceptions that still permit asbestos use are China, Russia, India, Brazil, Canada, and the United States. Replacing asbestos is a very hard task. In brake and clutch composites, asbestos has been especially important due to its heat resistance and low cost. Synthetic aramid, wollastonite (calcium silicate), mica (aluminum silicate), basalt fiber, cellulose, thermoset plastics, glass, and steel fibers are some common substitute products for asbestos in friction materials [33, 34].

**Figure 7.2** Asbestos fibers seen through an optical microscope.

**Figure 7.3** Six types of asbestos in its natural form.

According to Panda [34], all substitutes for asbestos have drawbacks, especially in high temperature applications, as in a brake system. Steel fibers, for instance, show good friction performance when used in brake friction materials. On the other hand, there are some reports showing that steel fibers are abrasive and can cause severe damage to the counterpart material (brake disk). Besides, steel is about four times as expensive as asbestos. Glass fiber is another well-known substitute for asbestos. Glass-fiber friction material can have a sudden loss in friction at high temperatures, since glass can melt. Moreover, there are health concerns about glass fibers, since they are suspected to cause skin irritations and lung cancer [35]. Finally, brake friction materials using glass and steel fibers tend to be noisier than those using asbestos [34].

Natural fibers (sisal, flax, jute, kenaf, coir, etc.) offer a number of advantages over synthetic and mineral fibers due to their renewable nature. Considering the high cost of aramid fibers and the environmental and health concerns related to the use of copper and asbestos, natural fibers become very attractive material candidates due to their low cost and environmental impact [36]. Cellulosic fibers, such as sisal, coconut fiber (coir), jute, palm, bamboo, wood, and paper, in their natural condition have been used as reinforcement agents of different polymer composites. They yield to composite biodegradability, since they are available from natural sources. Besides, vegetable fibers can provide significant acoustic insulation or damping properties to composite materials. Compared to mineral fibers (e.g. glass fiber), cellulosic fibers are able to provide to the composite certain benefits, such as low density; no risks to health; non-irritation to the skin; and a high degree of flexibility [37].

Mineral fibers have a brittle behavior. For this reason, they undergo breakage, while natural fibers tend to bend rather than fracture during processing [38, 39].

Thermal degradation, moisture absorption as well as degradation by microorganisms and sunlight (ultraviolet rays) are among the main issues of natural fiber reinforced composites [35]. Other drawbacks of natural fibers compared to synthetic fibers are their non-uniformity and variety of dimensions [9]. In the case of brakes, thermal degradation can limit the use of natural fibers in brake pads and linings, since the temperature can reach up to 750 °C during severe braking operations. Moisture absorption is related to the mechanical properties of the composite. According to the literature [38], the hydrophilic nature of vegetable fibers can lead to poor interfacial bonding between the polymer and the fiber [40].

A good interfacial bonding is required to achieve optimum reinforcement, and ultimately a good mechanical performance of the composite [40]. As explained in [41], the hydrophilic nature of natural fibers decreases the interfacial contact between them and the matrix polymer. The researchers added that water absorption may reduce the durability of the composites. Brake friction materials are required to be durable and resistant to impact and high shearing forces. To be certified by government agencies worldwide, brake friction materials are subjected to standard tests such as compressive strain test and impact test in order to ensure that they have good mechanical properties. Physical and chemical treatments have been used by many researchers as a way to improve the interfacial bonding of the natural fiber in the polymer matrix. It is, however, important to understand that fiber treatments (physical or chemical) have an important effect on the cost of the composite reinforced with natural fibers [41]. Finally, the third important issue related to natural fibers (degradation by microorganisms and sunlight) is also an important factor to consider when one plans to use it in brake friction materials. According to Ming et al. [41], the hydrophilic character of natural fibers provides the conditions for microorganisms to thrive. It is worth noting that brake materials are exposed to a variety of harsh environments, which include hot sunlight, rain, dust, etc.

Idris et al. [42] studied the tribological performance of banana peel fibers added to friction material composites. A pin-on-disk test apparatus was used to investigate the dry sliding wear rate of the samples in accordance with ASTM:G99-05 standard. This apparatus uses a white cast iron disk (hardness = 62 HRC), 120 mm track in diameter and 8 mm thick. Two sets of samples were produced with uncarbonized (BUNCp) and carbonized (BCp) banana peel particles (Figure 7.4). To form banana peel ash (carbonized, BCp), the raw banana peel powder was placed in an electric resistance furnace at a temperature of 1200 °C. Each set of samples (carbonized and uncarbonized) had six different wt% of

(a) (b)

**Figure 7.4** Photograph of banana peel powder: uncarbonized (BUNCp, (a)) and carbonized (BCp, (b)). *Source:* Idris et al. 2015 [42]. Reproduced with permission from Elsevier.

banana peel powder addition: 5%, 10%, 15%, 20%, 25%, and 30%. The resin used to bond the natural fibers was phenol formaldehyde. None of the commercial type of brake pads was included in this comparison.

In the study conducted by Idris et al. [42], the pin was pressed against the rotating white cast iron disk with a surface roughness of 0.3 μm at a load of 20 N and sliding speed of $5.02\,\mathrm{m\,s^{-1}}$. The total sliding distance of the tests was 5000 m. Table 7.1 summarizes the materials and test procedures used in the referred investigation.

As reported by Idris et al. [42], a uniform distribution of banana peels and the resin was obtained. The results of the tests showed that wear rate was reduced as the weight fraction of banana peels (wt%) increased for both BUNCp and BCp. The lowest wear rate (the best wear resistance) was found for 30 wt% of banana peel addition (BCp). The researchers attributed this behavior to high hardness values and compressive strength of the samples as the banana peel addition increased. Besides, BUNCp exhibited better wear resistance and slightly higher COF than BCp for all wt% of banana peel. On the other hand, the COF of the samples increased as the wt% of banana peel increased in the formulation. According to the researchers, the friction coefficient measured in the tests fell within the industrial standard range of 0.30–0.45 for automotive brake pads. Finally, the authors of the study concluded that samples containing 25 wt% in BUNCp and 30 wt% BCp showed the best overall properties. Also, they reported that banana peel particles can be used as an effective alternative to replace asbestos in the brake pad.

Yun et al. [36] compared the tribological performance of two regular brake friction materials (a commercial and a low-metallic, namely sample A and B, respectively) with three eco-friendly samples of brake materials (samples: C, D, and E). The eco-friendly materials were prepared by replacing a fraction or all of the metals and aramid with a natural fiber in a model brake lining formulation. The natural fibers added in the eco-friendly samples were provided by Interfibe (trademark: JMM). The authors of the study did not provide any property about those fibers. A brake dynamometer and the SAE recommended J2430 procedure were used to characterize the friction and wear performance of the brake samples. Furthermore, the brake effectiveness evaluation procedure (BEEP) criteria were used to assess the overall performance of brake pads during the J2430 procedure test. BEEP offers acceptance criteria in terms of friction effectiveness and structural integrity. Table 7.2 summarizes the test procedure, criteria assess-

**Table 7.1** Test materials and procedures used in [42].

| Disk roughness | Load | Sliding speed | Sliding distance | Standard |
|---|---|---|---|---|
| 0.3 μm | 20 N | $5.02\,\mathrm{m\,s^{-1}}$ | 5000 m | ASTM-G99-05 |
| **Materials type** | | **wt% banana peel** | | |
| Uncarbonized banana peel (BUNCp) | | 5, 10, 15, 20, 25, and 30 | | |
| Carbonized banana peel (BCp) | | | | |
| Commercial brake pad | | — | | |

**Table 7.2** Test procedure, BEEP criteria assessment, and material types used in [36].

| Standards | Assessments performed | | | | | |
|---|---|---|---|---|---|---|
| Test procedure SAE J2430 | Burnish and reburnish | Effectiveness | Fade | Heat performance | Cooling | Recovery |
| Criteria assessment BEEP | Effectiveness space | Cold effectiveness | Fade snubs | Heat performance #1 | Heat performance #1 | Structural integrity |
| **Materials type** | | **Aramid** | **Metals (copper, iron, tin, etc.)** | **Natural fiber (JMM)** | **Coke** | **Cost (\$ $kg^{-1}$)** |
| Sample A (commercial type) | | Not provided | | | | |
| Sample B (low-metallic brake pad) | | 3 | 8 | 0 | 16.8 | 1.381 |
| Sample C (eco-friendly) | | 3 | 0 | 3 | 16.8 | 1.149 |
| Sample D (eco-friendly) | | 0 | 0 | 6 | 16.8 | 0.826 |
| Sample E (eco-friendly) | | 0 | 0 | 6 | 14.2 | 0.839 |

ment, and materials used in the study of Yun et al. [36]. It is worth noting that the cost of each sample (in \$ $kg^{-1}$) is also shown in the table.

Results of the work conducted by Yun et al. [36] revealed that the commercial pad (sample A) exhibited lower speed sensitivity than other samples. But all samples (including commercial pad) showed decrease in COF as the speed was increased. Sample B exhibited a significant loss in COF during the fade section, while samples A and E showed an excellent fade resistance at high temperatures. Considering the brake performance, the eco-friendly and low-cost sample E showed a comparable performance to that of commercial brake pad (sample A). Both samples A and E demonstrated relative stability in COF, high fade resistance, good recovery capacity, and low sensitivity to speed. Sample A showed slightly better heat performance than sample E, but sample E slightly outperformed sample A when speed sensitivity was addressed. According to the authors, the better heat performance of sample A compared to that of sample E is mostly attributed to a difference in metal content. Compared with sample E (without metal content), sample A with a high metal content (8%) can better resist elevated temperatures. Only sample A and E passed in all BEEP criteria, while the other three samples B–D failed in heat performance, mainly due to their poor thermal resistance. Sample E is also a cost-effective brake pad. It is the second cheapest in the ranking.

Finally, the researchers suggested that the newly developed eco-friendly brake pad (sample E), which contains only natural fibers and developed without any hazardous material (e.g. copper, lead, etc.), can easily replace the current brake pads.

Ikpambese et al. [43] evaluated the friction and wear performance of automotive brake pads produced from palm kernel fibers (Figure 7.5). They compared their results with that of a commercial brake pad. The authors based their choice on earlier studies (e.g. [44, 45]), which showed that palm kernel fibers are a natural alternative to replace the hazardous ingredients used in brake pads.

Ikpambese et al. [43] developed six samples of brake pads with different percentages of palm kernel fibers and other ingredients (Table 7.3). The authors did

**Figure 7.5** Fibers of palm kernel used in [43]. *Source:* Ikpambese et al. 2016 [43]. Reproduced with permission from Elsevier.

**Table 7.3** Test materials and procedures used in [43].

| Brake disk material | Load | Sliding speed ($m\,s^{-1}$) | Sliding distance | Standard procedure |
|---|---|---|---|---|
| Cast iron disk | 25 000 N | 5.5,8.3,11.11, 16.6, 22.2, and 27.8 | — | — |

| | | Brake additives (wt%) | | | |
|---|---|---|---|---|---|
| Materials type | Epoxy resin | Palm kernel fibers | Aluminum oxide | Graphite | Calcium carbonate |
| Sample S1 | 15 | 35 | 5 | 5 | 40 |
| Sample S2 | 19 | 6 | 0 | 5 | 70 |
| Sample S3 | 23 | 27 | 10 | 10 | 30 |
| Sample S4 | 25 | 30 | 5 | 5 | 35 |
| Sample S5 | 30 | 40 | 5 | 5 | 20 |
| Sample S6 | 40 | 10 | 6 | 29 | 15 |
| Commercial brake pad | Not provided | | | | |

not provide any data about the formulation of the commercial brake pad used in the comparison. Each pair of brake pads developed was separately mounted on a brake dynamometer. The tests were performed at six sliding speeds (5.5, 8.3, 11.11, 16.6, 22.2, and 27.8 $m\,s^{-1}$) and under a fixed normal force of 25 000 N. The researchers did not follow any specific standard procedure in the tests.

As reported in [43], samples S1 (3.62 $mg\,m^{-1}$) and S3 (3.64 $mg\,m^{-1}$) exhibited lower wear rate than the commercial brake pad (3.8 $mg\,m^{-1}$). The lowest values of wear rate could be attributed to a good bonding of the natural fibers in the case of samples S1 and S3. S6 showed approximately the same wear rate as the commercial brake pad used in the authors' study. The others samples (S2, S4, and S5) exhibited a higher wear rate compared to the commercial friction material. All samples produced with palm kernel fibers exhibited a stable COF. The authors reported that this may be due to non-inclusion of steel fibers, which are responsible for making the COF unstable. Ikpambese et al. [43] also mentioned that values of COF for samples S1 (0.40) and S6 (0.33) agreed with the value range of 0.3–0.4 reported earlier by El-Tayeb and Liew [46].

About the effects of sliding speed on tribological properties, it was observed that the wear rates increased as speed increased for both the samples produced (S1 to S6) and the commercial brake pad. Any significant change in COF was observed for the whole range of sliding speed used in the study of Ikpambese et al. [43].

About the brake noise level, Ikpambese et al. [43] reported that most of the brake pads produced exhibited lower sound pressure level compared to the commercial brake pad. They attributed this behavior to non-inclusion of harder particles in the formulation of the produced samples. Sample S6 exhibited the best balance of tribological properties, i.e. high and stable friction coefficient along with a reduced wear rate and low brake noise. Finally, the authors concluded that palm kernel fibers can be effectively used as a replacement for hazardous ingredients in brake friction materials.

Bajpai et al. [47] investigated three different types of natural fibers (nettle, grewia optiva, and sisal), which were incorporated into plastic lactic acid (PLA) polymer to develop a new composite. Wear and frictional characteristics of the developed composites were investigated under dry contact condition at different operating parameters, such as applied load (10, 20, and 30 N), sliding speed (1, 2, and 3 $m\,s^{-1}$), and a fixed sliding distance (3000 m). As explained by the researchers, polylactic acid is a polymer derived from the fermentation of corn, potato, sugar beet, and other agriculture based materials. For this reason, it is considered a biopolymer. The composites were developed using the hot compression method. The dry sliding wear test on composite specimens has been carried out against a steel counterface using pin-on-disk apparatus (Ducom India; TR20LE). The procedure followed for the wear tests was ASTM G99-95 standard. Table 7.4 summarizes the operating conditions and materials used in the experiments conducted by Bajpai et al. [47].

Bajpai et al. [47] observed that COF decreased as both the applied load and speed increased for neat PLA and its composites. However, for all samples, COF appeared to be much more sensitive to the variation of applied normal load than the sliding speed variation. The researchers also noted that the overall average

Table 7.4 Test materials and procedures used in [47].

| Brake disk material | Load (N) | Sliding speed ($m\,s^{-1}$) | Sliding distance (m) | Standard procedure |
|---|---|---|---|---|
| Steel | 10, 20, and 30 | 1, 2, and 3 | 3000 | ASTM G99-95 |

| | Additives | |
|---|---|---|
| Materials type | Fiber | Fiber weight fraction (%) |
| Net polylactic acid | None | 0 |
| Polylactic acid/nettle | Nettle | 20 |
| Polylactic acid/grewia optiva | Grewia optiva | 20 |
| Polylactic acid/sisal | Sisal | 20 |

friction coefficient of PLA polymer was not much affected after natural fiber reinforcement was added. On the other hand, wear performance of PLA was significantly increased due to the addition of natural fibers. In other words, the wear resistance of the natural composites was considerably improved compared to the wear resistance of neat PLA, especially at higher loads. Results also showed that the wear rate of the developed composites is much more sensitive to the variation of applied normal load than the sliding speed variation. The authors of the study concluded that the natural fibers are a plausible candidate for reinforcement in composites for tribological applications. However, they did not report which sample exhibited the best tribological performance, nor whether the newly developed composites can be used as brake pads in brake applications.

Recent studies [41] have shown that hemp is taking a center stage as a source of suitable natural plant cellulose fibers because natural hemp bast fibers are long and inherently possess high strength. Compared to synthetic materials (e.g. glass fiber), hemp fibers have other advantages, such as low cost and low density, together with their high stiffness and strength-to-weight ratios. As explained in [41], hemp bast fibers seem to be the most advantageous natural fiber also because (i) hemp is an extremely fast growing crop, yielding more fibers per acre than other crops; (ii) hemp suppresses weeds effectively and is usually free from diseases and pests; and (iii) hemp leaves the soil in excellent condition for the succeeding crops.

Lee and Filip [10] studied the friction and wear performance of two eco-friendly automotive brake friction materials. These brake pads were manufactured with hemp fibers, which replaced two hazardous materials, copper and $Sb_2S_3$. The eco-friendly brake pads and a commercial (baseline) material were put to test in the full scale automotive brake dynamometer (Link Engineering model 2584) using a SAE J2430 test procedure. Besides, the BEEP criteria were used to assess the overall performance of brake pads during the J2430 procedure test. Table 7.5 summarizes the test materials and procedures used by Lee and Filip [10].

Results of the researchers' study (Lee and Filip [10]) showed that all tested brake pads passed the BEEP acceptance criteria. The dynamometer results showed that the modified samples had better performance when compared to the T-baseline as the temperature of the brake increased in the fade section of the SAE J2430 test. The sample T303 has the highest average effectiveness of 0.41 when compared to the samples T-baseline and T403, which exhibited values of 0.32 and 0.33, respectively. However, the modified samples showed higher wear rate than the T-baseline. The T-baseline had a thickness loss of 0.37 mm, but the T303 and T403 samples lost 1.43 and 2.36 mm in thickness, respectively. The researchers attributed the higher resistance of T-baseline to its better capacity to form a friction layer on the surface. As reported by the authors, T-baseline is covered with a fully developed and stable friction layer (third body) consisting mostly of Fe oxides, different formats of carbon, and compounds of materials originally present in the bulk brake material. This plays a considerable role when lowering the wear of brake materials. The T303 and T403 samples, formulated without Cu and Sb, did not develop a sufficient friction layer.

Mustafa et al. [9] used a weighted decision matrix method to compare different natural fibers to be used as an alternative source of friction material. The fibers compared in the authors' study were kenaf (the stems of plants of the family of malvaceae called *Hibiscus cannabinus*), jute, ramie, and asbestos. They listed the design criteria used in the decision matrix. As explained in [9], their decision matrix weighted four major design criteria, aimed at developing an eco-friendly lightweight automotive friction material. The four criteria were environmental friendliness (toxicity, $CO_2$ footprint, and safe for disposal), cost (price), performance (strength, stiffness, maximum service temperature, durability with water), and lightweight (density). The authors ranked the natural fibers from the highest to the lowest score, as follows: kenaf fibers (score: 8.74), jute fibers (score: 7.24), ramie fiber (score: 6.17), and asbestos (score: 4.26). Kenaf scored best in two criteria: density and toxicity (score 10 in both). On the other hand, jute and ramie fiber only scored 10 in toxicity, meaning they are both nontoxic. Asbestos, however, scored 10 in strength and 0 in toxicity. Table 7.6 summarizes the decision matrix made in [9].

Finally, the authors identified kenaf fibers as the most suitable material to be used in brake friction materials, since it passes all the design requirements. The authors concluded that their results show promising potential for kenaf fibers by their capability in eco-awareness with low impact to the environment.

Xin et al. [48] studied the friction and wear characteristics of brake composites reinforced with different contents of sisal fibers. As explained by the researchers, sisal fiber is a kind of plant fiber with cellulose structure. A piece of sisal fiber is not a single filament like carbon or glass fiber. It consists of

**Table 7.5** Test materials and procedures used in Lee and Filip [10].

| Standards | Assessments performed | | | | | | |
|---|---|---|---|---|---|---|---|
| Test procedure: SAE J2430 | | Burnish and ruburnishing | Effectiveness | Fade | Heat performance | Cooling | Recovery |
| Criteria assessment: BEEP | | Effectiveness space | Cold effectiveness | Fade snubs | Heat performance #1 | Heat performance #1 | Structural integrity |
| Materials type | Copper (wt%) | $Sb_2S_3$ (wt%) | Phenolic resin (wt%) | Geopolymer (wt%) | Kelvar fiber (wt%) | Hemp fiber (wt%) | Others (lubricants, abrasive, filler, ceramic) (wt%) |
| T-baseline (commercial type) | 8 | 3.6 | 9.5 | 0 | 3.4 | 0 | 75.5 |
| T 30-3 (eco-friendly) | 0 | 0 | 6.6 | 2.9 | 1.7 | 1.7 | 87.1 |
| T 40-3 (eco-friendly) | 0 | 0 | 5.7 | 3.8 | 1.7 | 1.7 | 87.1 |

**Table 7.7** Test materials and procedures used in [48].

| Brake disk material | Pressure (MPa) | Sliding speed ($m\,s^{-1}$) | Duration (min) | Standard procedure |
|---|---|---|---|---|
| **Gray cast iron** | **0.98** | **2.00** | **10** | - |

| | | | Additives | | | | | |
|---|---|---|---|---|---|---|---|---|
| **Study** | **Goal** | **Materials type** | **Resin** | **Sisal fiber** | **Copper** | **Barite** | **Felspar** | **Others (ZnO, friction powers, clay)** |
| Stage I | Identify the optimal wt% of sisal fiber | S1 | 15 | 10 | 10 | 15 | 10 | 40 |
| | | S2 | 15 | 15 | 10 | 15 | 10 | 35 |
| | | S3 | 15 | 20 | 10 | 15 | 10 | 30 |
| | | S4 | 15 | 25 | 10 | 15 | 10 | 25 |
| | | S5 | 15 | 30 | 10 | 15 | 10 | 20 |
| Stage II | Compare the optimal sample (S3) with commercial brake pads (C1, C2, and C3) | C1 (asbestos fiber) | Not provided | | | | | |
| | | C2 (glass fiber) | Not provided | | | | | |
| | | C3 (steel and mineral fiber) | Not provided | | | | | |
| | | S3 (sisal fiber, 20 wt%) | 15 | 20 | 10 | 15 | 10 | 30 |

**Table 7.8** Summary of the main results on the use of natural fibers in brake materials described in this chapter.

| | Reference | | | | | | |
|---|---|---|---|---|---|---|---|
| | **Ikpambese et al. [43]** | **Yun et al. [36]** | **Idris et al. [42]** | | **Xin et al. [48]** | **Lee and Filip [10]** | **Bajpai et al. [47]** |
| Natural material | Palm kernel | Interfiber JMM (a cellulose fiber) | Carbonized banana peel (BCp) | Uncarbonized banana peel (BUNCp) | Sisal | Hemp fiber | Sisal, nettle, and grewia optiva pin-on-disk newly developed |
| Classification | Fiber | Fiber/ reinforcement | Binder | Binder | Fiber | Fiber | Fiber/reinforcement |
| Other ingredients | Epoxy resin, aluminum oxide, graphite, and calcium carbonate | Phenolic resin, fillers, friction modifiers, reinforcement | Not provided | Not provided | Resin copper, barite, kevlar, ZnO, felspar, and others | Phenolic resin, geopolymer, kevlar, and others | Polymer polylactic acid PLA |
| Standard procedure | Not provided | SAE J2430 | ASTM-G99-05 | ASTM-G99-05 | Constant speed tester | SAE J2430 | Pin-on-disc ASTM:G99-05 |
| Temperature behavior | — | — | Not provided | Not provided | COF of sisal fiber brake composites decreased at temperatures higher than 250 °C | Not provided | Not provided |
| Friction behavior | COF was stable in the whole range of speed | Effectiveness of all eco-friendly samples decreased with increasing speed | ↑ wt% ↑ COF | ↑ wt % COF | From 5 to 20 wt% ↑ wt% ↑ COF<br>From 25 to 30 wt% ↑ wt% ↓ COF | COF was between 0.32–0.41 for all samples of eco-friendly brake pads | For all samples, COF decreases as applied load or speed increases. Samples are much more sensitive to the variation of applied normal load than the sliding speed variation |

*(Continued)*

**Table 7.8** (Continued)

| | Reference | | | | | | |
|---|---|---|---|---|---|---|---|
| | **Ikpambese et al. [43]** | **Yun et al. [36]** | **Idris et al. [42]** | | **Xin et al. [48]** | **Lee and Filip [10]** | **Bajpai et al. [47]** |
| Wear resistance | ++++ | ++++ | ++ | +++ | +++ | ++++ | + |
| Fade resistance | Not provided | ++++ | Not provided | Not provided | Not provided | +++ | Not provided |
| Recovery capacity | Good recovery capacity | Good recovery capacity | Not provided | Not provided | Not provided | Not provided | Not provided |
| Optimal composition | 40 wt% epoxy-resin, 10 wt% palm kernel fiber, 6 wt% $Al_2O_3$, 29 wt% graphite, and 15 wt% calcium carbonate | 6 wt% JMM fiber, 0 wt% aramid and metals, 14.2 wt% of coke, others (79.8 wt%) | 30 wt% | 25 wt% | Sisal fiber in 20 wt%, 15 wt% resin, 10 wt% copper, 15 wt% barite, 10 wt% felspar, and 30 wt% others | 5.7–6.6 wt% of phenolic resin, 2.9–5.7 wt% geopolymer, 1.7 wt% kevlar, others (87 wt%) | Not provided |
| Remarks | The optimal composition exhibited high and stable friction coefficient along with a reduced wear rate and low brake noise. The authors concluded that palm kernel fibers can be used in commercial brake pads | Eco-friendly composite passed in all BEEP criteria. It is also a cost-effective brake pad | Although fade resistance and recovery capacity have not been tested, authors suggest that banana peels can be used in brake pads | | The optimal composite of sisal fiber showed the same COF as commercial brake pads. Meanwhile, it exhibited a higher wear rate than the commercial brake pad. But it is still in the eligible range | All eco-friendly samples of brake pads have passed the brake effectiveness evaluation procedure (BEEP) | The researchers did not inform whether the new developed composites can be used as brake pads in brake applications |

## References


1 Blau, P.J. (2001). Compositions, functions, and testing of friction brake materials and their additives. Technical Report. Oak Ridge National Laboratory, Oak Ridge, USA.

2 Eriksson, M., Bergman, F., and Jacobson, S. (2002). On the nature of tribological contact in automotive brakes. *Wear* 252: 26–36.

3 Eriksson, M. and Jacobson, S. (2000). Tribological surfaces of organic brake pads. *Tribology International* 33: 817–827.

4 Ostermeyer, G.P. (2001). Friction and wear of brake systems. *Forschung im Ingenieurwesen* 66: 267–272.

5 Ostermeyer, G.P. (2003). On the dynamics of the friction coefficient. *Wear* 254: 852–858.

6 Eriksson, M. (2000). Friction and contact phenomena of disc brakes related to squeal. Thesis. Acta Universitatis Upsaliensis, Uppsala, Sweden.

7 Rhee, S. (1990). The role of friction film in friction, wear, and noise of automotive brakes. SAE paper number 900004.

8 Ostermeyer, G.P. and Müller, M. (2008). New insights into the tribology of brake systems. *Journal of Automotive Engineering*, Special Issue on Automotive Braking 222 (D7): 1167–1200.

9 Mustafa, A., Abdollah, M.F.B., Shuhimi, F.F. et al. (2015). Selection and verification of kenaf fibres as an alternative friction material using weighted decision matrix method. *Materials and Design* 67: 577–582.

10 Lee, P.W. and Filip, P. (2013). Friction and wear of cu-free and SB-free environmental friendly automotive brake materials. *Wear* 302 (1–2): 1404–1413.

11 Innovate US (2013). The copper on brakepads can pollute the water. www.innovateus.net/earth-matters/copper-brakepads-can-pollute-water (accessed July 2017).

12 Sandahl, J.F., Baldwin, D.H., Jenkins, J.J., and Scholz, N.L. (2007). A sensory system at the interface between urban stormwater runoff and salmon survival. *Environmental Science and Technology* 41 (8): 2998–3004.

13 Rhee, S.K. and Schwartz, H.W. (1979). Test methods for automotive and truck friction materials. *American Society for Testing and Materials* 39–49, paper 701.

14 Rusnak, R.M., Spurgeon, W.M., and Aldrich, F.W. (1970). Friction material testing. Society of Automotive Engineers, paper 700517.

15 Tsang, P.H.S., Jacko, M.G., and Rhee, S.K. (1985). Comparison of chase and inertial brake dynamometer testing of automotive friction materials. *Wear* 103 (3): 217–232.

16 Müller, M.B. (2009). Proposta de uma metodologia para desenvolvimento de novo fornecedor de freios traseiros a tambor para veículos já em produção. Master Dissertation (in Portuguese), 86 pages. Escola Politécnica da Universidade de São Paulo, Brazil.

17 Zimmer, D. and Teves, A. (1982). ATE friction test machine and other methods of lining screening. SAE.

18 Haynes Publishing (1999). Automotive disc brake manual. Techbook Series.

19 Papinniemi, A., Stanef, D., Moore, S., and Hamdi, S. (2007). Brake noise in practice. 14th International Congress on Sound and Vibration (ICSV 14), Australia, 8 pages.

20 Blau, P.J. and McLaughlin, J.C. (2003). Effects of water films and sliding speed on the frictional behavior of truck disc brake materials. *Tribology International* 36: 709–715.

21 Cristol-Bulthé, A., Desplanques, Y., Degallaix, G., and Berthier, Y. (2007). Mechanical and chemical investigation of the temperature influence on the tribological mechanisms occurring in OMC/cast iron friction contact. *Wear* 264: 815–825.

22 Desplanques, Y., Roussette, O., Degallaix, G. et al. (2007). Analysis of tribological behaviour of pad-disc contact in railway braking part 1. Laboratory test development, compromises between actual and simulated tribological triplets. *Wear* 262 (5–6): 582–591.

23 Lee, K., Blau, P.J., and Truhan, J.J. (2007, 2007). Effects of moisture adsorption on laboratory wear measurements of brake friction materials. *Wear* 262: 925–930.

24 Kumar, M. and Bijwe, J. (2010). Studies on reduced scale tribometer to investigate the effects of metal additives on friction coefficient, temperature sensitivity in brake materials. *Wear* 269: 838–846.

25 Neis, P.D., Ferreira, N.F., Sukuraman, J. et al. (2015). Characterization of surface morphology and its correlation with friction performance of brake pads. *International Journal Sustainable Construction and Design* 6 (2): 6.

26 Limpert, R. (1999). *Brake Design and Safety*, 2e. SAE International.

27 Amador, A., Flynn, B., and Ward, D. (2004). Friction and wear of materials. Technical Report. Rensselaer Polytechnic Institute, New York, USA.

28 Neis, P.D., Ferreira, N.F., and Lorini, F.J. (2011). Contribution to perform high temperature tests (fading) on a laboratory-scale Tribometer. *Wear (Lausanne)* 271: 2660–2664.

29 John, M.J. and Thomas, S. (2012). *Natural Polymers Volume 1: Composites*. RSC Publishing.

30 Alleman, J.E. and Mossman, B.T. (1997). Asbestos revisited. *Scientific American* 277: 54–57.

31 IARC Monograph (2012). Asbestos (chrysotile, amosite, crocidolite, tremolite, actinolite, and anthophyllite), vol. 100C. http://monographs.iarc.fr/ENG/Monographs/vol100C/mono100C-11.pdf (accessed July 2017).

32 Asbestos Nation (2017). Asbestos bans around the world. www.asbestosnation.org (accessed September 2017).

33 Nicholson, G. (1995). *Facts about Friction*. Croydon: P&W Price Enterprises, Inc.

34 Panda, H. (2008). *The Complete Technology Book on Asbestos, Cement, Ceramic and Limestone*. Asia Pacific Business Press Inc.

35 Ali, A., Shaker, K., Nawab, Y. et al. (2016). Hydrophobic treatment of natural fibers and their composites – a review. *Journal of Industrial Textiles* 2153–2183. doi: 10.1177/1528083716654468.

36 Yun, R., Filip, P., and Lu, Y. (2010). Performance and evaluation of eco-friendly brake friction materials. *Tribology International* 43: 2010–2019.

37 Girisha, C., Sanjeevamurthy, and Rangasrinivas, G. (2012). Tensile properties of natural fiber-reinforced epoxy hybrid composites. *International Journal of Modern Engineering Research* 2 (2): 471–474.

38 Herrera-Franco, P.J. and Valadez-Gonzalez, A. (2004). Mechanical properties of continuous natural fibre-reinforced polymer composites. *Composites: Part A* 35: 339–345.

39 Herrera-Franco, P.J. and Valadez-Gonzalez, A. (2005). A study of the mechanical properties of short natural-fiber reinforced composites. *Composites: Part B* 36: 597–608.

40 Pickering, K.L., Aruan Efendy, M.G., and Le, T.M. (2016). A review of recent developments in natural fibre composites and their mechanical performance. *Composites Part A: Applied Science and Manufacturing* 83: 98–112.

41 Ming, L., Thygesen, A., Summerscales, J., and Meyer, A.S. (2017). Targeted pre-treatment of hemp bast fibres for optimal performance in biocomposite materials: a review. *Industrial Crops & Products* 108: 660–683.

42 Idris, U.D., Aigbodion, V.S., Abubakar, I.J., and Nwoye, C.I. (2015). Eco-friendly asbestos free brake-pad: using banana peels. *Journal of King Saud University – Engineering Sciences* 27 (2): 185–192.

43 Ikpambese, K.K., Gundu, D.T., and Tuleun, L.T. (2016). Evaluation of palm kernel fibers (PKFs) for production of asbestos-free automotive brake pads. *Journal of King Saud University – Engineering Sciences* 28: 110–118.

44 Ibhadode, A.O.A. and Dagwa, I.M. (2008). Development of asbestos-free friction lining material from palm kernel shell. *Journal of the Brazilian Society of Mechanical Sciences and Engineering* 30 (2): 166–173.

45 Koya, O.A. and Fono, T.R. (2010). Palm kernel shell in the manufacture of automotive brake pad. www.rmrdctechnoexpo.com (accessed September 2017).

46 El-Tayeb, N.S.M. and Liew, K.W. (2009). On the dry and wet sliding performance of potentially new frictional brake pad materials for automotive industry. *Wear* 266: 275–287.

47 Bajpai, P.K., Singh, I., and Madaan, J. (2013). Tribological behavior of natural fiber reinforced PLA composites. *Wear* 297: 829–840.

48 Xin, X., Guang Xu, C., and Qing, L.F. (2007). Friction properties of sisal fibre reinforced resin brake composites. *Wear* 262: 736–741.

# 8

# Comparative Wear Model on Hybrid Natural Fiber Composites as Substitutions for UHMWPE Made Knee Implants

*Gusztáv Fekete and Mátyás Andó*

*Eötvös Loránd University, Faculty of Informatics, Savaria Institute of Technology, Szombathely H-9700, Hungary*

## 8.1 Introduction

### 8.1.1 Basics of Reinforced Polymers, Composites, and Their Testing

Engineering materials have undergone a great evolution in history. Depending on the technology available at a given time, certain preferences were made for materials as well. This relative importance of materials is seen in Figure 8.1. As is seen, due to the work and success of material researchers, more and more new polymer and composites materials have been developed in the last decades.

Besides polymers, hundreds of new materials exist with regard to ceramics or metals. As a result, traditional classifications – metal, polymer, ceramic, and composites – changed to application based classifications (aerospace materials, electronic materials, smart materials, etc.). To understand hybrid natural fiber composites (NFCs) better, a brief introduction about polymers is necessary.

### 8.1.2 Classification of Polymers

The industrial history of polymers started with rubber in the twentieth century. The first spurt in the application of different polymers was around the time of the Second World War. For the last 20–30 years the applications and the types of polymers increased continuously. It must be noted that polymers can be synthetic or natural, but this classification only refers to the source, not to the structure.

Polymers are organic materials that have specific macromolecular structures. Monomers are the basic units, which can repeat 10 000 times in a single polymer chain. The simplest monomer is ethylene ($C_2H_4$), which has double bonds between carbon atoms.

The following chemical process has to be carried out in order to create a simple polymer chain: the double bonds connecting two neighboring monomers must be broken first, and a new single bond connection must be created (Figure 8.2) to achieve a chain.

*Synthesis and Tribological Applications of Hybrid Materials*, First Edition.
Edited by Mohammad Jawaid, Rajini Nagarajan, Jacob Sukumaran, and Patrick De Baets.

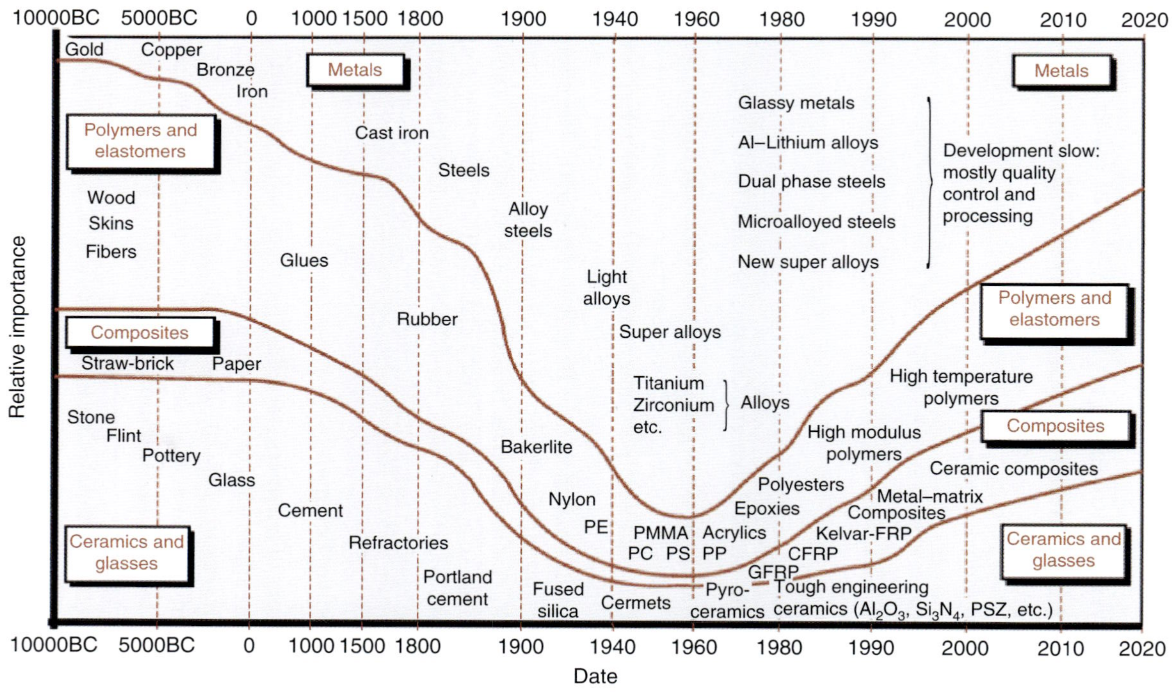


**Figure 8.1** The relative importance of the materials in the history. *Source:* Ashby 2009 [1]. Reproduced with permission from Elsevier.

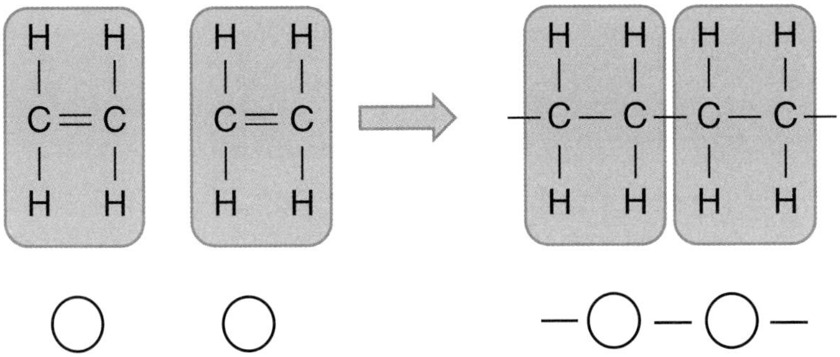


**Figure 8.2** Ethylene monomers before and after the reaction and their symbolic representation.

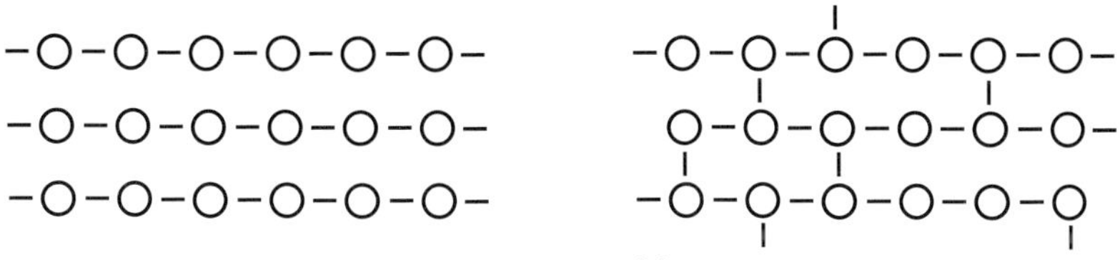


**Figure 8.3** Structures of thermoplastics (a) and thermosets (b).

There are thousands of monomers that create different macromolecular structures as well. Based on these structures the polymers can be thermoplastics or thermosets. The basic difference between the two structures is the cross-link. In the case of thermoplastics there are no cross-links between the chains (Figure 8.3a); therefore, these materials can remold after solidification. The thermosets have cross-links; thus during the chemical reaction when they are formed, they cannot be remolded any more (Figure 8.3b).

In thermoplastic structures, there are only secondary chemical bonds between the polymer chains. If the temperature rises to a limited point, the secondary chemical bonds break and the polymer chains can slip next to each other. Primary chemical bonds need much more energy to break so that the polymer chains could remain complete. After cooling, the new position of the chains becomes fixed by the new secondary chemical bond. Thermosets cannot melt since there are only primary chemical bonds in their structure. At much higher temperatures, both the chain and the cross-link can be broken.

Naturally, in this elevated temperature the thermoplastic is also degraded, but there is still a technological range in the temperature where they can melt and a new shape can be created (by injection molding, extrusion, etc.).

Thermoplastic polymers have two different subtypes in their structures. If the molecules solidify in a random arrangement, then the polymers become amorphous (Figure 8.4a). Amorphous thermoplastics are generally transparent. If polymer chains show some dimensional order locally, then they are called semicrystalline polymers (Figure 8.4b).

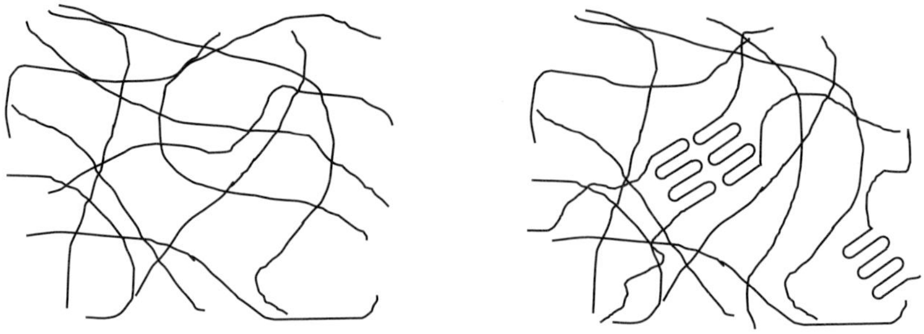


**Figure 8.4** Structure of amorphous (a) and semi-crystalline polymers (b).

The degree of crystallinity is one of the important parameters that have direct connection to the material property of the polymer. The degree of crystallinity can vary from 1% to 100%.

Thermoset polymers have cross-links between the chains that provide better heat resistance. Based on the amount of cross-links two groups can be differentiated: slightly cross-linked thermosets or elastomers (e.g. rubber) and highly cross-linked thermosets or duromers (e.g. epoxy resin).

### 8.1.3 Classification of Composites

Composites have different definitions. Nowadays, if a material has at least two additives, then companies in material science business term it a "composite," but mostly this is only for marketing reasons. The polymer industry uses the so-called fillers since the beginning of material production. Originally, if a filler can reinforce the polymer and improve its mechanical performance, then it can be a composite. In the case of polymers, the most typical reinforcing material is probably glass fiber, and it is widely used in thermoplastics and thermosets as well.

Fibers mostly have improved mechanical behaviors than the polymer (matrix). The purpose of the fiber is to make stronger connection between two neighboring areas in the material. Here, something must be made clear. In engineering, Young's modulus is likely the most important parameter, rather than tensile strength. If the material has high modulus, then the resistance of the deformation is also high (high stiffness). This consideration is the main theory behind reinforced composites. Fibers also have the role of transferring load if they have appropriate connection with the matrix. The fibers connect to the matrix through chemical adhesion. If the adhesion is strong enough, then the composite will have adequate stiffness as well. If the adhesion is low, the fibers tend to loosen (pull out) and they create a dramatic decrease in the stiffness (Figure 8.5).

If the matrix has only one fiber reinforcement, the maximum effect can only be achieved if the fiber breaks (tensile strength – $\sigma$). This also means that the adhesion must be strong enough, which depends on the surface and the shear strength ($\tau$) of the fiber–matrix connection. Based on a mechanical equation (if the fiber has perfect cylindrical surface) the critical length ($L_c$) can be calculated:

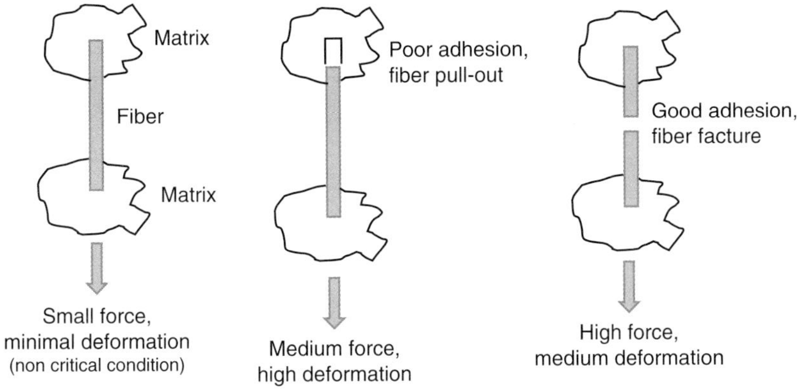


**Figure 8.5** Composite behavior in case of different adhesion.

$$L_c = D\frac{\sigma_f}{\tau} \tag{8.1}$$

If the fiber is longer than the critical length, then it works effectively. In many cases, it is easier to achieve shorter fibers, which also makes working with it easier (e.g. injection molding), than enhancing the chemical adhesion to a higher level. There are different treatments to achieve better chemical connections that can improve the shear strength between the fiber and the matrix.

If the fiber is long enough and the deformation remains in the linear region of the material, then Hooke's law can be applied with regard to the parallel fibers:

$$E_c = E_f V_f + E_m (1 - V_f) \tag{8.2}$$

where $E_c$ is the Young's modulus of the composites, $E_f$ is the Young's modulus of the fiber, $E_m$ is the Young's modulus of the matrix, and $V_f$ is the volume of the fiber. To achieve real reinforcement by applying fibers, usually the Young's modulus and the tensile strength must be kept relatively high while the density is low. Table 8.1 includes some typical fibers used for reinforcement and their properties [2].

**Table 8.1** Typical fibers for reinforcement use.

| Fiber type | Density ($g\,cm^{-3}$) | Tensile strength (GPa) | Young's modulus (GPa) |
|---|---|---|---|
| Steel | 7.86 | 4.0 | 210 |
| Glass fiber (E-type) | 2.6 | 2.5 | 72 |
| Carbon fiber (HS) | 1.78 | 3.4 | 240 |
| Aramid (Kevlar 49) | 1.44 | 3.3 | 75 |
| Polyethylene (SK 66) | 0.97 | 3.3 | 99 |

**Table 8.2** Typical natural fibers.

| Fiber type | Density ($g\,cm^{-3}$) | Tensile strength (GPa) | Young's modulus (GPa) |
|---|---|---|---|
| Bamboo (manual extraction) | 0.91 | 0.5 | 35.9 |
| Palm | 1.03 | 0.4 | 2.7 |
| Coconut | 1.15 | 0.5 | 2.5 |
| Banana | 1.35 | 0.6 | 17.9 |
| Sisal | 1.45 | 0.6 | 10.4 |

Different plants or animals form the source of natural fibers. It is generally true that these fibers are biodegradable and they are from renewable sources. Table 8.2 lists some typical natural fibers and their properties [3].

Composite materials can be classified according to their length and their orientation. Based on the length, there are two types: short fiber/discontinuous fiber (longer than the critical length) and continuous fiber (c. thousand times longer than the critical length). The orientations are also of two types: random (fibers have different directions) and aligned (all the fibers are in the same direction).

In two dimensions (sheet forms), there can be continuous aligned structures (Figure 8.6a), short aligned structures (Figure 8.6b), short random structures (Figure 8.6c), and continuous random structures (Figure 8.6d). In addition, different fiber weaves/fabrics also exist (Figure 8.6e).

Composite structures can be more complex in three dimensions. Depending on the technologies, usually there are more layers laid on each other. 2D aligned structures become random structures if the different layers have free directions.

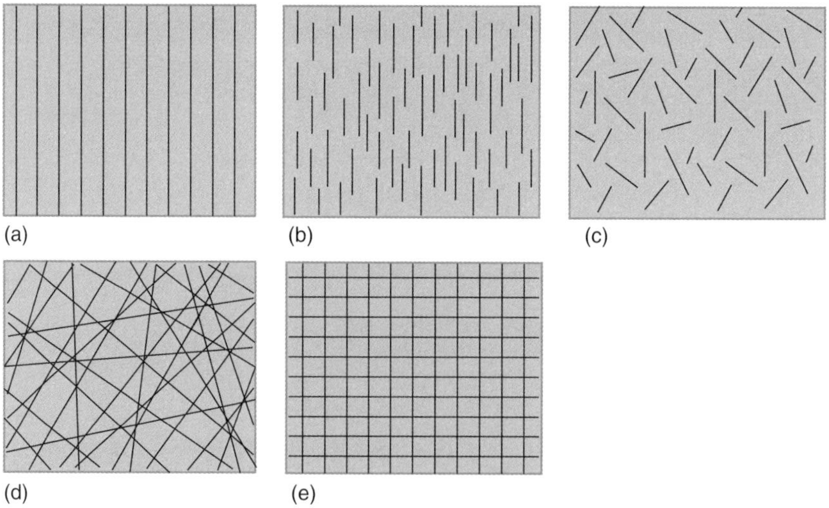


**Figure 8.6** (a–e) Different composite structures in two dimensions.

It must be noted that 3D fabrics exist that have oriented structure in 3D as well. These types of structures allow engineers to design composite parts where the reinforcement completely fits the stress distribution and the directions.

Additional need for optimizations and special properties led the researchers in composite technology to create innovative solutions. Hybrid composites are an answer to this need. Hybrid composites contain at least two types of fibers, where the different fibers can be in the same sheet (2D), or the different layers can create a hybrid system in 3D.

### 8.1.4 Basics of Tribo-testing

Tribology is a science that deals with friction, wear, and lubrication. In reality, two interacting surfaces create a complex phenomenon with regard to tribology. In order to obtain a clear view about this contact phenomenon in relation to tribological behaviors, a schematic tribological system was created (Figure 8.7).

Basically, all tribological systems include two bodies, the motion of which can be described with the law of relative kinematics. Usually, the intermediate material plays the role of a lubricant and the wear debris. The rest of the parameters (e.g. temperature, load, sliding distance, surface roughness, materials) belong to the tribological environment. During an experiment, the researchers attempt to identify the main parameters and to be able to control them in the future experiments. One of the basic equations in tribology is the ratio of the coefficient of friction:

$$\mu = \frac{F_F}{F_N} \tag{8.3}$$

where $F_F$ is the friction force and $F_N$ is the normal force. Both the friction force and the normal force are influenced by many parameters that are included in the tribological environment. Owing to these parameters and their statistical deviance, high deviations are expected during repeated tribological measurements, even though the parameters are seemingly fixed. Figure 8.8 contains the classification of tribo tests.

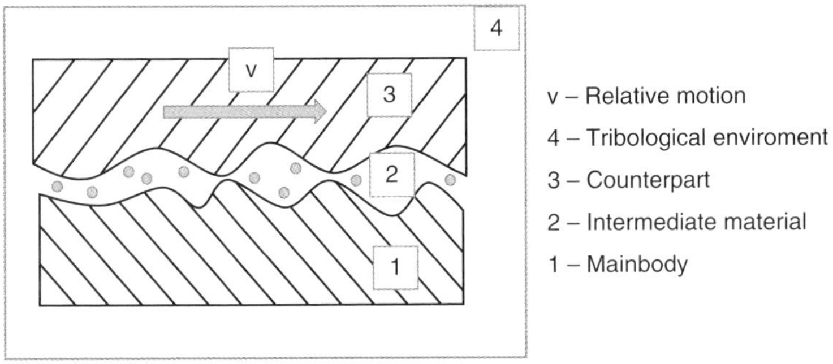


**Figure 8.7** Simplified tribological system.

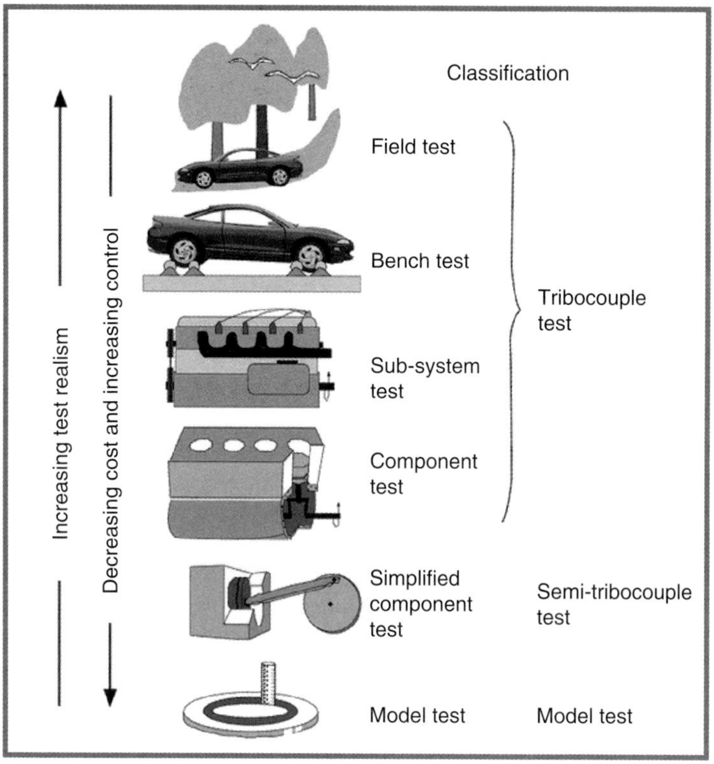


**Figure 8.8** Classifications of tribo tests. *Source:* Bhushan 2001 [4]. Reproduced with permission from CRC Press.

Tests can be performed in almost endless number of ways. Typical factors are running time, accurate control of test conditions, isolated wear mechanisms, and so on. Generally, a test model is used to better identify the main tribological behavior. Testing a simplified model is considerably easier, quicker, and cheaper under well-controlled test conditions, but it also correlates less with reality. Creating an appropriate model to address a real tribological problem is probably the most challenging part of the research. Researchers need to deal with basic effects (approximately the seven more important ones) to simulate the tribological conditions.

These effects are as follows:

1) Contact (tribocouples: ball on plate, block on ring, block on plate, pin on disk, etc.)
2) Load (forces, Hertzian contact, etc.)
3) Displacement (sliding, rolling, spinning, flowing, oscillating, impact, etc.)
4) Relative motion (constant, periodical, magnitude, etc.)
5) Track (clean surface, transfer film, lubrication, debris, etc.)
6) Surface roughness ($R_a$, $R_z$, etc.)
7) Temperature

Based on the earlier results, typical wear and friction models have been already established. The principle wear mechanisms are recognized as adhesive wear, abrasive wear, erosive wear, corrosive wear, and fatigue wear. Mostly, the wear debris changes the tribological environment by altering its complexity and deviations. In addition, in real applications there are no clear conditions that bring forth more complex wear phenomena.

### 8.1.5 Hybrid Natural Fiber Composites and Their Possible Use in Total Knee Replacements (TKR)

As was mentioned earlier, fibers are mostly classified based on their source, which can be related to minerals, plants, and animals. Biologically, these materials are made up of either cellulose or proteins depending on the source. Lately, there has been a rapid growth in R&D in the NFC area.

The particular interest in this material is due to the advantages in their mechanical properties compared to synthetic fiber composites, not mentioning the low environmental impact cost as well. It must be also mentioned that usually higher performance can be credited to plant fibers than the available animal fibers with regard to their higher strengths and stiffness.

Among the numerous possible applications, companies involved in biomedical part production are also interested in this new type of materials as a possible substitution for UHMWPE manufactured elements. Naturally, these new materials can only enter the market if they are able to provide acceptable wear characteristics and resilience compared to the materials being used today.

Undeniably, problems related to wear in total knee replacements (TKRs) have decreased in the past 10 years [5] but not disappeared. In the design-related factors, wear is still the second most important mechanical factor (first is the periprosthetic fracture) that limits the lifetime of TKRs and it also greatly depends on the local kinematics of the knee.

On studying the wear properties of these materials, it becomes quite evident that NFCs have significant difference in their wear behavior. An important parameter that is widely used in analytical and numerical modeling is the specific wear coefficient (denoted as $k$), which is summarized in Table 8.3.

It must be also noted that these materials were tested approximately with the same sliding speed, but the load is significantly less than in the study by Patten

**Table 8.3** Specific wear coefficient in case of natural fibers.

| Author | Material | Load (N) | Velocity ($mm\,s^{-1}$) | Specific wear coefficient ($mm^3\,N^{-1}\,m^{-1}$) |
|---|---|---|---|---|
| Chittaranjan and Acharya [6] | *Lantana camara* | 20 | 31.4 | $1.5\times10^{-11}$ |
| Ren et al. [7] | Hybrid PTFE | 20 | 26 | $1.4\times10^{-14}$ |
| Omrani et al. [8] | Jute | 20 | 30 | $4\times10^{-7}$ |
| Yousif et al. [9] | Palm fiber reinforced polyester | 20 | 30 | $2.5\times10^{-6}$ |
| Patten et al. [10] | UHMWPE against CoCr | 200 | 30 | $1.3\times10^{-6}$ |

et al. [10], where the authors tested materials that are actually being used in TKR production.

By looking at these results, it is apparent that the specific wear coefficient of the *Lantana camara* and the hybrid PTFE are exceptional, but it is unknown yet whether they could perform as well as the UHMWPE if the load is elevated not only to 200 N but significantly beyond. The palm fiber reinforced polyester is in the same range regarding the specific wear coefficient compared to the standard UHMWPE materials used in TKRs.

Since the load in the TKR can be up the six to seven times the body weight (BW) during squatting and two to three BW during gait, more experimental data are needed before these materials could enter the market not only as substitutional choices but also as prime material possibilities.

## 8.2 Aims

In this chapter, the main aim is to introduce an enhanced wear model that can be a useful tool for estimating wear outcome for NFCs, if the specific wear coefficient is available. As a first step, a classical UHMWPE material is chosen to give an insight on how the analytical wear model works if knee-specific parameters such as the slide–roll ratio or the varying load in case of non-standard and standard squat are incorporated into it.

As for wear in knee implants, it must be noted that this phenomenon is, on one hand, caused by the non-congruent form of the knee joint together with its natural instability. On the other hand, emerging particle debris can also be a relevant cause of cartilage damage. Nevertheless, it must be pointed out that the phenomenon of wear is in relation with several interrelated factors; thus, it must be examined as a system, not as a material property [11].

Wear is the second among the main mechanical factors that limit the lifetime duration of TKRs [5, 12, 13] and it is also highly dependent on the kinematics of the knee joint [14]. Beside multiple other parameters in wear studies, slide–roll ratio between 0% and 40% is frequently applied during tribological tests carried out with pin-on-disk test rigs or knee simulators [15, 16]. During wear tests, its magnitude is always considered constant based on the study of McGloughlin and Kavanagh [17] and Hollman et al. [18]. This approximation is correct if it is applied on pin-on-disk or ball-on-disk tests, where the kinematics is precisely adjusted to this special configuration.

Nevertheless, several authors published results that are against the idea of constant slide–roll ratio regarding TKRs [19, 20] where they referred to the complex geometry, which highly alters the local kinematics. It was demonstrated in the study of Laurent et al. [21] that the wear mechanism is highly dependent not only on the load in the contact but also on the interfacial contact kinematics, which includes a cyclic multidirectional path of motion and the slide–roll ratio.

Thus, varying the slide–roll ratio under the different motions (gait, squat, etc.) can be a possible parameter to give a better approximation on TKR wear mechanism.

In the following, a more realistic mathematical approach on wear propagation between the tibiofemoral connections will be introduced. For the analysis, a commonly applied wear model, Archard's equation, is used [22], which incorporates the friction force, derived from the tibiofemoral force in the contact, with a varying slide–roll ratio.

In spite of its simplicity, Archard's equation is widely used in the contemporary literature [23, 24] and the predicted wear is in reasonable agreement with other more sophisticated models such as the model of Turel et al. [25] or the model of Abdelgaied et al. [26]. Squat is chosen for analyses regarding wear, since during this motion the highest contact forces [27] and the highest slide–roll ratio can be approached [20]. Along with the abovementioned fact, squat is a widely exercised movement for rehabilitation and for thigh muscles strengthening due to sport or medical reasons [28, 29]; therefore, it is important to know the aspect of wear with regard to this motion.

## 8.3 Methods

### 8.3.1 Wear Modeling

The instantaneous volume of material removed from the tibiofemoral surface of the TKR due to mild wear is predicted by Archard's law [22]:

$$\mathrm{d}V = k F_{\mathrm{N}}\, \mathrm{d}s \tag{8.4}$$

where $k$ is the specific wear rate ($\mathrm{mm}^3\ \mathrm{N}^{-1}\ \mathrm{m}^{-1}$), which is a material-dependent quantity; $F_{\mathrm{N}}$ is the normal force acting between the contact surfaces, which in this case is the contact surface of the tibia and the femur (N); and d$s$ is the infinitesimal length of sliding (m). There are two parameters in this equation that are to be further enhanced: the force and the sliding length wherein the slide–roll ratio can be defined.

The wear mechanism between the connecting surfaces is supposed to be abrasion based (two-body). This means that during contact, the harder metallic femoral part ploughs into the softer polyethylene tibial part, and in the connecting point(s) reaction force(s) appear (Figure 8.9). As a first step, we can approximate these forces with the Coulomb law:

$$F_{\mathrm{friction}} = \mu_{\mathrm{k}} F_{\mathrm{N}} \tag{8.5}$$

where $\mu_{\mathrm{k}}$ is the coefficient of kinetic friction.

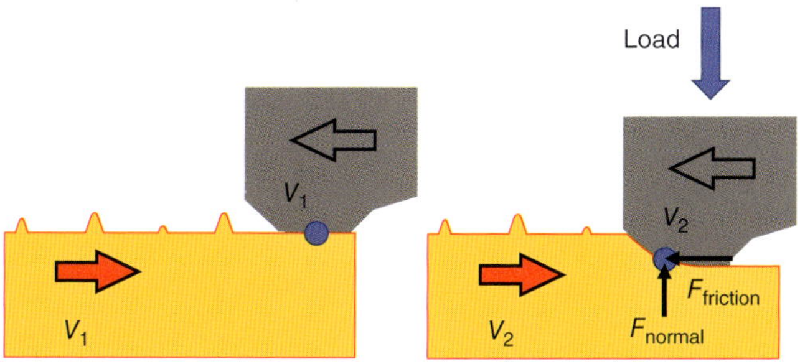


**Figure 8.9** Two-body abrasive wear between the connecting surfaces.

In the abrasive wear mechanism, the frictional component is responsible for creating such a shear stress in the upper surface of the material that it begins to lose small debris. Therefore, it provides us a more precise approximation if the friction force, deduced from the tibiofemoral force, is introduced in our wear equation:

$$dV = k F_{\text{friction}}\, ds = k \mu_k F_N\, ds \tag{8.6}$$

### 8.3.2 Force Modeling for Wear Equation

The normal force in Equation (8.6) that acts between the connecting surfaces is considered constant during wear tests. Nevertheless, in case of the tibiofemoral connection during squat movement, this force is a flexion angle dependent quantity and is called tibiofemoral force. It has been deduced analytically for standard (no horizontal movement of the center of gravity is permitted) and non-standard [27] (horizontal movement of the center of gravity is allowed as it is in real life) squat motion (Figure 8.10), which makes it an adequate choice to incorporate

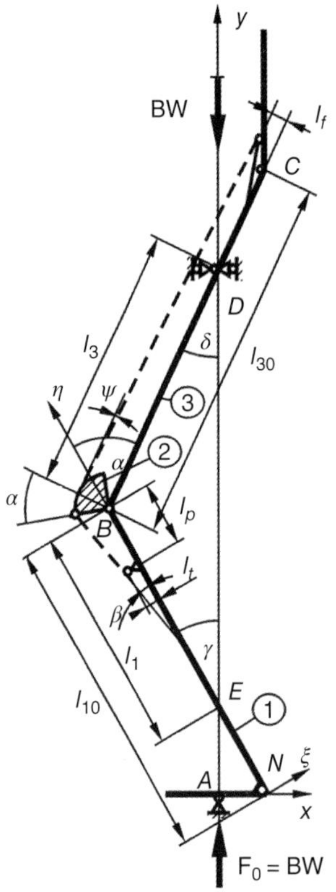


Figure 8.10 Mechanical model of squat with horizontally moving center of gravity.

this force into our model. The function of the tibiofemoral force can be described as follows [27]:

$$\frac{F_{\mathrm{tf}}(\alpha)}{\mathrm{BW}} = \frac{\lambda_1(\alpha)\sin\gamma(\alpha)}{\lambda_{\mathrm{p}}\sin\beta(\alpha) + \lambda_{\mathrm{t}}\cos\beta(\alpha)}\left(\frac{\cos\beta(\alpha)}{\cos\varphi(\alpha)} + \frac{\cos\gamma(\alpha)}{\cos\varphi(\alpha)}\right) \tag{8.7}$$

while the $\varphi(\alpha)$-function is described as [27]

$$\varphi(\alpha) = \mathrm{arctg}\left[\frac{(\lambda_1(\alpha) - \lambda_{\mathrm{p}})\mathrm{tg}\beta(\alpha) - \lambda_{\mathrm{t}}}{\lambda_1(\alpha)\cdot tg\gamma(\alpha) + \lambda_{\mathrm{p}}\mathrm{tg}\beta(\alpha) + \lambda_{\mathrm{t}}}\mathrm{tg}\gamma(\alpha)\right] \tag{8.8}$$

The parameters of the equation are summarized in Table 8.4.

If Equation (8.7) is multiplied by an arbitrarily chosen BW then the tibiofemoral force is obtained. In the abovementioned analytical expressions, $\lambda_1$ function describes the horizontal movement of the center of gravity during squat. If the linear function is used as given in Table 8.4, then a more realistic, so-called non-standard squat is carried out where the torso leans forward, and it helps the knee to release a considerable amount of moment (Figure 8.11). This is the squat type that is practiced by most people in everyday life.

The so-called standard squat is defined for tests and simple mechanical modeling. Nonetheless, it does not consider the movement of the torso during the motion; therefore, higher knee moment, tibio- and patellofemoral forces act in the knee if this motion is considered. With this model, both motions, and therefore the two types of tibiofemoral forces, are analyzed with their impact on wear.

When the non-standard squat is being examined the $\lambda_1$ function is constant with a value of 1. This means that the torso does not lean forward (Figure 8.11, left drawing). In case the non-standard squat is exercised, the torso leans forward, which is modeled by a flexion angle dependent $\lambda_1(\alpha)$ function (Figure 8.11, right drawing).

In the augmented wear equation, the tibiofemoral force will be included as follows:

$$\mathrm{d}V = kF_{\mathrm{friction}}\,\mathrm{d}s = k\mu_{\mathrm{k}}F_{\mathrm{tf}}(\alpha)\mathrm{d}s \tag{8.9}$$

**Table 8.4** Parameters of the tibiofemoral force and $\varphi$ function [27].

| Parameter | Quantity or function | SD |
|---|---|---|
| $\lambda_1(\alpha)$ | $0.0024{\cdot}\alpha + 0.492$ (–) | 0.15 |
| $\beta(\alpha)$ | $-0.3861{\cdot}\alpha + 26.56$ (°) | 14 |
| $\o(\alpha)(\o = \gamma/\alpha)$ | $-0.0026{\cdot}\alpha + 0.567$ (–) | 0.081 |
| $\lambda_{\mathrm{t}}$ | 0.11 (–) | 0.018 |
| $\lambda_{\mathrm{p}}$ | 0.1475 (–) | 0.043 |

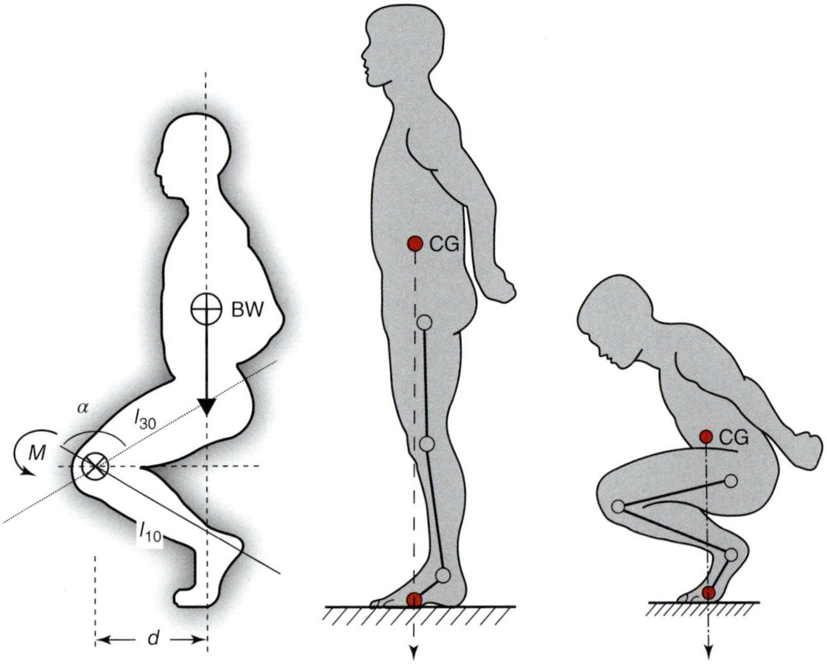


**Figure 8.11** Standard (non-moving CoG) and non-standard (moving CoG) squatting.

If the tibiofemoral force is indexed as $F_{\text{tf-nst CoG}}$ then the horizontal movement of the center of gravity is also included, while with the index of $F_{\text{tf-st CoG}}$, the forward lean of the torso, the horizontal motion of the center of gravity, is excluded.

### 8.3.3 Slide–Roll Modeling for Wear Equation

Slide–roll is known to be an important parameter of wear; however, its quantitative effect has not been studied earlier. By knowing the slide–roll function, exact conclusions can be drawn about the features of the motion. If the ratio is 0 then pure rolling is present, while 1 describes pure sliding. If the ratio is between 0 and 1, the movement is characterized as partial rolling and sliding. For example, a sliding–rolling ratio of 0.4 means 40% of sliding and 60% of rolling. Fekete et al. [20] introduced the slide–roll ratio by having the arc lengths on both connecting bodies as follows:

$$\mathrm{S/R}(t) = \frac{\Delta s_{\text{tibia}}(t) - \Delta s_{\text{femur}}(t)}{\Delta s_{\text{tibia}}(t)} \tag{8.10}$$

where

$$\Delta s_{\text{femur}}(t) = s_{\text{femur}}(t) - s_{\text{femur}}(t - \Delta t)$$

$$\Delta s_{\text{tibia}}(t) = s_{\text{tibia}}(t) - s_{\text{tibia}}(t - \Delta t)$$

are the corresponding incremental differences of the connecting arc lengths.

A positive ratio shows the slip of the femur compared to the tibia. By this definition, the author derived an averaged slide–roll ratio with a standard deviation (SD = 0.136) based on several TKRs as a function of flexion angle [20]:

$$S/R(\alpha) = -5.16\times10^{-7}\alpha^3 + 1.235\times10^{-4}\alpha^2 - 4.113\times10^{-3}\alpha + 0.226 \quad (8.11)$$

This function is used further on in the calculations, and it is assumed that the function is applicable for flexion and extension movement as well. With respect to the slide–roll ratio, there is an accordance between the authors that sliding is not dominant up to 65–67° of flexion angle [17, 18], which practically represents the domain of gait swing.

Nägerl et al. [19] and Fekete et al. [20] derived lower ratios between 0 and 65–67 flexion angle (Figure 8.12), and above this region they presume that the slide–roll ratio can reach 95% of sliding as well [19]. The interpretation for why these ratios have such discrepancies is likely to be credited to the simplifications, e.g. frictionless contact, two-dimensional, simple circle or simplified knee geometry that the earlier authors applied in their models [17, 18]. Nägerl et al. [19] used the first-time three-dimensional geometry to derive the slide–roll ratio, while Fekete et al. [20] included the effect of friction besides three dimensionality in their three-dimensional models.

Owing to these enhancements, the latter two authors revealed a nonlinear phenomenon in the trend of the slide–roll ratio. Since slide–roll is known, it must be involved in the wear calculation to quantify its effect. To include the

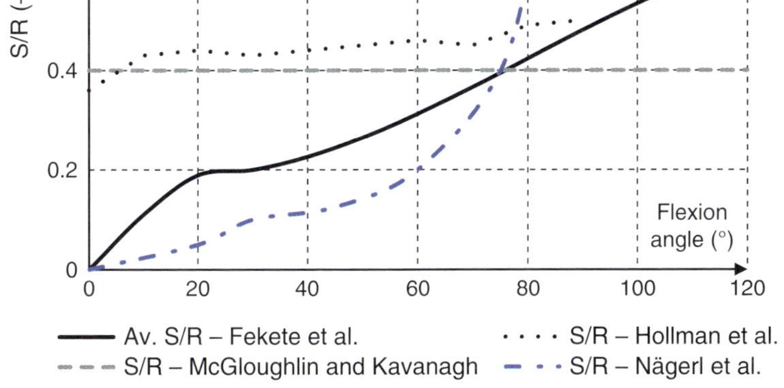


**Figure 8.12** Summarized slide–roll ratios from different authors.

slide–roll ratio into the equation, let us rewrite the length of sliding as the product of sliding velocity ($\mathrm{m\,s^{-1}}$) and infinitesimal time (s).

$$\mathrm{d}V = k\mu_k F_{tf}(\alpha)\mathrm{d}s = k\mu_k F_{tf}(\alpha)v_{sliding}(\alpha)\mathrm{d}t \tag{8.12}$$

The definition of slide–roll according to Fekete et al. [20] can be expressed by the tangential velocities instead of the arc lengths:

$$\mathrm{S/R}(t) = \frac{v_{CTt}(t) - v_{CFt}(t)}{v_{CTt}(t)} = \frac{v_{sliding}(t)}{v_{CTt}(t)} \rightarrow v_{sliding}(t) = \mathrm{S/R}(t)\cdot v_{CTt}(t) \tag{8.13}$$

where $v_{CTt}$ and $v_{CFt}$ are the tangential velocities in the instantaneous contact points. The difference between these velocities ($v_{CTt} - v_{CFt} = v_{sliding}$) provides the so-called sliding velocity. If Equation (8.13) is rearranged to $v_{sliding}$ then it can be included in our equation and we obtain the augmented Archard's law:

$$\mathrm{d}V = k\mu_k F_{tf}(\alpha)v_{CTt}(t)\mathrm{S/R}(t)\mathrm{d}t \tag{8.14}$$

$v_{CTt}$ velocity is kept constant similar to a wear test parameter [10] as an approximation. The original function of $F_{tf}$ is the function of flexion angle [20], but it can be transformed into time domain. This has been carried out from the study of Fekete et al. [20], where the data were obtained also in time domain during the squat simulation. Thus, the final equation is as follows:

$$\mathrm{d}V = k\mu_k \mathrm{BW}\, F_{tf}(t)v_{CTt}\,\mathrm{S/R}(t)\mathrm{d}t \tag{8.15}$$

The parameters for the equation are summarized in Table 8.5.

**Table 8.5** Wear parameters.

| Parameter | Quantity or function |
| --- | --- |
| $k$: Specific wear rate [10] | $1.3 \times 10^{-6}$ ($\mathrm{mm^3\,N^{-1}\,m^{-1}}$) |
| $\mu_k$: Coefficient of kinetic friction [20] | 0.003 (–) |
| $F_{tf}$/BW: Tibiofemoral force relative to the body weight (moving CoG) | $0.770\,2{\cdot}t^3 + 0.554{\cdot}t^2 + 1.624\,4{\cdot}t + 1.031\,1$ (–) |
| $F_{tf}$/BW: Tibiofemoral force relative to the body weight (non-moving CoG) | $0.665\,2{\cdot}t^3 + 0.439\,7{\cdot}t^2 + 3.262{\cdot}t + 1.056\,4$ (–) |
| S/R: varying slide–roll ratio | $-0.111\,667{\cdot}t^3 + 0.363\,173\,2{\cdot}t^2 + 0.023\,783\,65{\cdot}t + 1.137\,558$ (–) |
| S/R: constant slide–roll ratio | 0.4 (–) |
| BW: body weight | 1000 (N) |
| $v_{CTt}$: Tangential velocity of the tibia in the contact | 30 ($\mathrm{mm\,s^{-1}}$) |
| $t$: Duration of motion | During squat: 1.53 (s) |

In the following, six total wear volumes will be calculated, each of them during one squat cycle. One cycle in squat is defined between 0° and 120° of flexion while the complete range where the propagation of wear monitored is 15 years. A usual knee wear simulation period is 3.5 million cycles [30, 31]; however, squat and deep squat are less frequently exercised in everyday life.

If 3.5 million cycles were broken down over 15 years, it would mean 640 complete deep squat cycles carried out daily. This number of cycles is clearly overestimated; thus 50 deep squats per day will be assumed in this chapter. The total wear during one cycle can be calculated by integrating the wear functions over time. During deep squat, the total cycle will be expanded to 120° of flexion angle where the time interval goes from 0 to 3.06 s during the squat motion (Figure 8.13).

The total wear volumes are calculated in the following way: firstly, the slide–roll ratio is kept constant with the number 0.4 according to the study of McGloughlin and Kavanagh [17] together with Hollmann et al. [18] while a standard squat is carried out (the tibiofemoral force does not have the effect of the horizontally moving center of gravity). Secondly, the slide–roll ratio varies according to the function of Fekete et al. [20] while a standard squat is carried out as well. Thirdly, the slide–roll ratio is kept constant with the number 0.4, but non-standard squat is considered (horizontally moving center of gravity), while fourthly, the slide–roll ratio varies during the non-standard squat.

In the last two calculations, similarly to a wear test on a pin-on-disk or ball-on-disk configuration, the slide–roll ratio is kept constant at 0.4 in one case, while in the other case it varies.

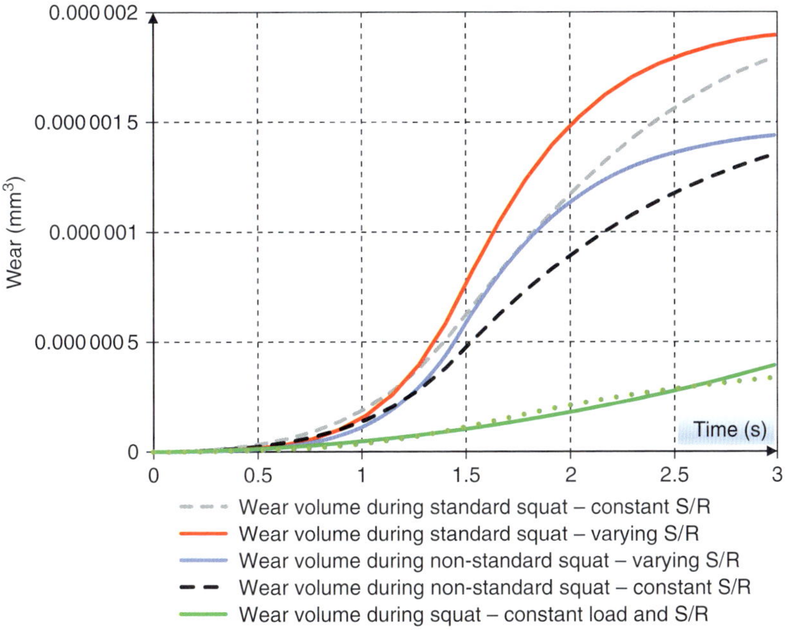


**Figure 8.13** Wear propagation of the different models.

Regarding the force, only the simple Coulomb law ($F_s = \mu_k F_N$) is considered, which results in a constant load. The general method to calculate total wear during one squat cycle is as follows:

$$\int_{V_0}^{V_{\text{total}}} dV = \int_{t_0}^{T} k\mu_k \, \text{BW}\, F_{\text{tf}}(t) v_{\text{CTt}}\, \text{S/R}(t) dt \tag{8.16}$$

$$V_{\text{total}} = \int_{t_0}^{T} k\mu_k \, \text{BW}\, F_{\text{tf}}(t) v_{\text{CTt}}\, \text{S/R}(t) dt + V_0 \tag{8.17}$$

In the first calculations, both models use the tibiofemoral force where the horizontal movement of the center of gravity is included (non-standard squat "nst"). Thus, the total wear with varying slide–roll ratio is

$$V_{\text{nst-vS/R}} = \int_{0}^{3.06} k\mu_k \, \text{BW}\, F_{\text{tf-nstCoG}}(t) v_{\text{CTt}}\, \text{S/R}(t) dt = 2.04 \times 10^{-6} \left(\text{mm}^3/\text{cycle}\right) \tag{8.18}$$

while the total wear with constant slide–roll ratio is

$$V_{\text{nst-cS/R}} = \int_{0}^{3.06} k\mu_k \, \text{BW}\, F_{\text{tf-nstCoG}}(t) v_{\text{CTt}}\, \text{S/R} dt = 1.76 \times 10^{-6} \left(\text{mm}^3/\text{cycle}\right) \tag{8.19}$$

In the second calculation, both models use the tibiofemoral force where the center of gravity does not move horizontally (standard squat "st").

Total wear with varying slide–roll ratio but without the horizontal movement of the center of gravity:

$$V_{\text{st-vS/R}} = \int_{0}^{3.06} k\mu_k \, \text{BW}\, F_{\text{tf-stCoG}}(t) v_{\text{CTt}}\, \text{S/R}(t) dt = 2.68 \times 10^{-6} \left(\text{mm}^3/\text{cycle}\right) \tag{8.20}$$

Total wear with constant slide–roll ratio:

$$V_{\text{st-cS/R}} = \int_{0}^{3.06} k\mu_k \, \text{BW}\, F_{\text{tf-stCoG}}(t) v_{\text{CTt}}\, \text{S/R} dt = 2.33 \times 10^{-6} \left(\text{mm}^3/\text{cycle}\right) \tag{8.21}$$

Total wear if constant load (BW) and constant slide–roll ratio are used:

$$V_{\text{cF+cS/R}} = \int_{0}^{3.06} k\mu_k \, \text{BW}\, v_{\text{CTt}}\, \text{S/R} dt = 0.4289 \times 10^{-6} \left(\text{mm}^3/\text{cycle}\right) \tag{8.22}$$

And, at last, total wear if constant load (BW) and variable slide–roll ratio are used:

$$V_{\text{cF+vS/R}} = \int_{0}^{3.06} k\mu_k \, \text{BW}\, v_{\text{CTt}}\, \text{S/R}(t) dt = 0.4295 \times 10^{-6} \left(\text{mm}^3/\text{cycle}\right) \tag{8.23}$$

## 8.4 Results

The calculations of the integrals have been summarized in Table 8.6, where the percentile difference of the new parameters can be observed. By looking at the calculation of the wear test condition, only 1% difference is observable when varying or constant slide–roll ratio is applied. Therefore, the number of $0.429 \times 10^{-6}$ (mm$^3$/cycle) can be chosen as a point of reference.

When the results from the standard and non-standard squat are compared to the reference value, it is visible that the wear, generated by the augmented models, is assumed to be 4.16–6.25 times more compared to the reference value.

The significant difference is due to the new parameters, the varying tibiofemoral force function, and the slide–roll ratio. As for quantitative conclusions, the total wear with constant slide–roll ratio during standard squat ($V_{\text{const S/R}}$) is expected to be 5.5 times higher compared to the reference value ($V_{\text{W test}}$), while if total wear includes varying slide–roll ratio during standard squat, then the calculated wear is ~6.25 times higher.

With regard to non-standard squat, total wear with constant slide–roll ratio during standard squat ($V_{\text{const S/R}}$) is expected to be 4.16 times higher compared to the reference value ($V_{\text{W test}}$), while if the total wear includes varying slide–roll, then the calculated wear is ~4.75 times higher.

In case the wear model is coupled with constant slide–roll ratio under non-standard squatting as compared to a wear model where the slide–roll ratio varies and standard squat is considered, then a 65% higher wear volume can be expected with respect to the model with standard squat movement.

It is easy to notice that the highest wear volume is generated in the case of standard squat due to the fact that the tibiofemoral force, without the effect of the horizontal lean of the torso, will reach the highest magnitude compared to other squatting motion. This makes the tibiofemoral force probably the most dominant parameter in the augmented Archard's law.

**Table 8.6** Wear results and comparison.

| Wear volume ($10^{-6}$ mm$^3$/cycle) | | Standard squat $\frac{V_{\text{const S/R}}}{V_{\text{W test}}}$ | Standard squat $\frac{V_{\text{vary S/R}}}{V_{\text{W test}}}$ | Non-standard squat $\frac{V_{\text{const S/R}}}{V_{\text{W test}}}$ | Non-standard squat $\frac{V_{\text{vary S/R}}}{V_{\text{W test}}}$ |
|---|---|---|---|---|---|
| Wear test condition | $V_{\text{cF+cS/R}} = 0.4289$ | | | | |
| | $V_{\text{cF+vS/R}} = 0.4295$ | | | | |
| Standard squat | $V_{\text{st-vS/R}} = 2.68$ | | | | |
| | $V_{\text{st-cS/R}} = 2.33$ | 5.5 | 6.25 | 4.16 | 4.75 |
| Non-standard squat | $V_{\text{nst-vS/R}} = 2.04$ | | | | |
| | $V_{\text{nst-cS/R}} = 1.76$ | | | | |

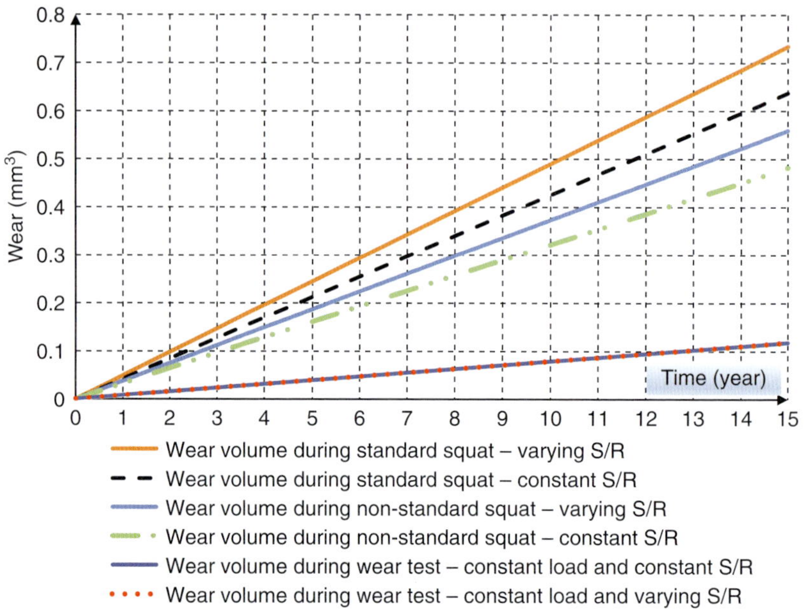


**Figure 8.14** Total wear during lifetime cycle.

According to Figure 8.13, the varying slide–roll ratio has also the effect of increasing the wear volume. A feasible explanation for why the wear volume is lower when constant 0.4 (40% of sliding) slide–roll ratio is considered during squat can be addressed as follows: it has been proved by several studies that during flexion of the knee joint, approximately up to 20–30° of flexion, rolling is dominant [19, 20], which is not a crucial motion as far as wear is concerned, since the friction effect caused by sliding is also very low.

Therefore, the authors who considered 0.3–0.4 slide–roll ratio were correct with respect to the beginning of the motion of squat, but not with respect to the complete cycle. Above 30° of flexion, as sliding prevails, wear starts developing between the surfaces due to shear stress, which is caused by the increasing slide [30]. Since the kinematic condition of the TKR geometries is different, wear must differ during the progression of the motion as well, which can be observed in Figure 8.13.

The propagation of wear for the lifetime cycle has been calculated as well, as seen in Figure 8.14.

Evidently, in case the wear test model is considered, where the load is constant different slide–roll ratio types (constant or varying) have negligible effect on wear in the long term.

## 8.5 Discussion

The effect of the new parameters on wear, the slide–roll ratio, and the tibiofemoral force with and without the horizontal movement of the center of gravity have been evidently demonstrated and quantitatively determined. Varying slide–roll ratio as a wear parameter causes approximately 15% more wear than constant slide–roll ratio, regarding any type of squatting motion. The involvement of the tibiofemoral force has a major effect on wear, which can lead to 65% more removed volume per cycle compared to other configurations.

As a further step, on one hand, the augmented wear model is to be applied for other types of movements, such as gait or even specific motions, e.g. cutting movement in football. This can be carried out by knowing the load case (load function) during these motions. On the other hand, if the specific wear coefficients of NFCs could be estimated with higher load cases as well, then the model would be appropriate to estimate wear propagation in the case where these materials played a role in a TKR configuration instead of a standard UMHWPE material.

Limitations of the study must be pointed out as well. The friction force is calculated by the simple Coulomb law, which assumes a point connection, while in reality the contact is a surface. The specific wear rate ($k$) can highly alter the phenomenon, and it can be only obtained via experiments for different kinematic conditions coupled with diverse materials. It should be studied further how well it is possible to estimate the correct $k$ value from the literature in order to gain adequate results for the mathematical analysis.

## Acknowledgments

This work was supported by the ELTE Eötvös Loránd University in the frame of the ÚNKP-17-4 New National Excellence Program of the Ministry of Human Capacities, the National Natural Science Foundation of China (81301600) and Anta Sports Products Limited (HK2015000090), the Savaria Institute of Technology, and the Research Academy of Grand Health, Ningbo University.

## References

1 Ashby, M.F. (2009). *Material Selection in Mechanical Design*. Elsevier.
2 Young, R.J. (2011). *Introduction to Polymers*, 3e. Taylor & Francis Group.
3 Rao, K.M.M. and Rao, K.M. (2007). Extraction and tensile properties of natural fibers: Vakka, date and bamboo. *Composite Structures* 77: 288–295.
4 Bhushan, B. (2001). *Modern Tribology Handbook*. CRC Press.
5 Sharkey, P.F., Lichstein, P.M., Shen, C. et al. (2014). Why are total knee arthroplasties failing today – has anything changed after 10 years? *Journal of Arthroplasty* 29: 1774–1778.
6 Chittaranjan, D. and Acharya, S.K. (2010). Effects on fiber content on abrasive wear of *Lantana camara* fiber reinforced polymer matrix composite. *Indian Journal of Engineering and Material Sciences* 17: 219–223.
7 Ren, G., Zhang, Z., Song, Y. et al. (2017). Effect of MWCNTs-GO hybrids on tribological performance of hybrid PTFE/Nomex fabric/phenolic composite. *Composite Science and Technology* 146: 155–160.
8 Omrani, E., Menezes, P.L., and Rohatgi, P.K. (2016). State of the art on tribological behavior of polymer matrix composites reinforced with natural fibers in the green materials world. *Engineering Science and Technology, an International Journal* 19 (2): 717–736.

9 Yousif, B.F. and El-Tayeb, N.S.M. (2009). Wet adhesive wear characteristics of untreated oil palm fibre-reinforced polyester and treated oil palm fibre-reinforced polyester composites using the pin-on-disc and lock-on-ring techniques. *Proceedings of the Institution of Mechanical Engineers, Part J: Journal of Engineering Tribology* 224: 123–131.
10 Patten, E.W., Van Citters, D., Ries, M.D., and Pruitt, L.A. (2013). Wear of UHMWPE from sliding, rolling, and rotation in a multidirectional tribo-system. *Wear* 304 (1–2): 60–66.
11 Karlhuber, M. (1995) Development of a method for the analysis of the wear of retrieved polyethylene components of total knee arthroplasty. Thesis. Technical University of Hamburg, Germany.
12 Guo, Y., Hao, Z., and Wan, C. (2016). Tribological characteristics of polyvinylpyrrolidone (PVP) as a lubrication additive for artificial knee joint. *Tribology International* 77: 214–219.
13 Qiu, M., Chyr, A., Sanders, A.P., and Raeymaekers, B. (2014). Designing prosthetic knee joints with bio-inspired bearing surfaces. *Tribology International* 77: 106–110.
14 Wimmer, M.A. and Andriacchi, T.P. (1997). Tractive forces during rolling motion of the knee: implications for wear in total knee replacement. *Journal of Biomechanics* 30 (2): 131–137.
15 Lopez-Cervantes, A., Dominguez-Lopez, I., Barceinas-Sanchez, J.D.O., and Garcia-Garcia, A.L. (2013). Effects of surface texturing on the performance of biocompatible UHMWPE as a bearing material during in vitro lubricated sliding/rolling motion. *Journal of the Mechanical Behavior of Biomedical Materials* 20: 45–53.
16 Rawal, B.R., Yadav, A., and Pare, V. (2016). Life estimation of knee joint prosthesis by combined effect of fatigue and wear. *Procedia Technology* 23: 60–67.
17 McGloughlin, T. and Kavanagh, A. (1998). The influence of slip ratios in contemporary TKR on the wear of ultra-high molecular weight polyethylene (UHMWPE): an experimental view. *Journal of Biomechanics* 31: 8.
18 Hollman, J.H., Deusinger, R.H., Van Dillen, L.R., and Matava, M.J. (2002). Knee joint movements in subjects without knee pathology and subjects with injured anterior cruciate ligaments. *Physical Therapy* 82: 960–972.
19 Nägerl, H., Frosch, K.H., Wachowski, M.M. et al. (2008). A novel total knee replacement by rolling articulating surfaces. In vivo functional measurements and tests. *Acta of Bioengineering and Biomechanics* 10 (1): 55–60.
20 Fekete, G., De Baets, P., Wahab, M.A. et al. (2012). Sliding–rolling ratio during deep squat with regard to different knee prostheses. *Acta Polytechnica Hungarica* 9 (5): 5–24.
21 Laurent, M.P.T., Johnson, S., Yao, J.Q. et al. (2003). In vitro lateral versus medial wear of knee prosthesis. *Wear* 255 (7–12): 1101–1106.
22 Archard, J.F. and Hirst, W. (1956). The wear of metals under unlubricated conditions. *Proceedings of the Royal Society A* 236: 397–410.
23 Pal, S., Haider, H., Laz, P.J. et al. (2008). Probabilistic computational modeling of total knee replacement wear. *Wear* 264 (7–8): 701–707.

24 O'Brien, S.T., Bohm, E.R., Petrak, M.J. et al. (2014). An energy dissipation and cross shear time dependent computational wear model for the analysis of polyethylene wear in total knee replacements. *Journal of Biomechanics* 47 (5): 1127–1133.
25 Turell, M., Wang, A., and Bellare, A. (2003). Quantification of the effect of cross-path motion on the wear rate of ultra-high molecular weight polyethylene. *Wear* 255 (7–12): 1034–1039.
26 Abdelgaied, A., Liu, F., Brockett, C. et al. (2011). Computational wear prediction of artificial knee joints based on a new wear law and formulation. *Journal of Biomechanics* 44 (6): 1108–1116.
27 Fekete, G., Csizmadia, B.M., Wahab, M.A. et al. (2014). Patellofemoral model of the knee joint under non-standard squatting. *Dyna Colombia* 81 (183): 60–67.
28 Slater, L.V. and Hart, J.M. (2017). Muscle activation patterns during different squat techniques. *Journal of Strength and Conditioning Research* 31 (3): 667–676.
29 Slater, L.V. and Hart, J.M. (2016). The influence of knee alignment on lower extremity kinetics during squats. *Journal of Electromyography and Kinesiology* 31: 96–103.
30 Liu, A., Jennings, L.M., Ingham, E., and Fisher, J. (2015). Tribology study of natural knee using and animal model in a new whole joint natural knee simulator. *Journal of Biomechanics* 48 (12): 3004–3011.
31 Brockett, C.L., Carbone, S., Fisher, J., and Jennings, L.M. (2017). PEEK and CFR-PEEK as alternative bearing materials to UHMWPE in a fixed bearing total knee replacement: an experimental wear study. *Wear* 374–375: 86–91.

# 9

# Fabrication and Tribological Behavior of Epoxy Hybrid Composites

*Bheemappa Suresha*[1] *and Rajashekaraiah Hemanth*[2]

[1] *The National Institute of Engineering, Department of Mechanical Engineering, Manandavadi Road, Mysore 570 008, Karnataka, India*
[2] *NIE Institute of Technology, Department of Mechanical Engineering, #50, Next to BEML, Koorgalli, Mysore 570 018, Karnataka, India*

## 9.1 Introduction

The components that are engaged in relative motion with their counterparts in any product demand synergy of mechanical and tribological behavior for service longevity. Traditional materials often demonstrate better performance in either of these, resulting in early failure of the part. This has provided the scope for evolution of tailor-made materials, i.e. composites, to suit the desired applications. Among the available composites, polymer matrix composites (PMCs) offer benefits – for instance, superior strength to weight ratio, resistance to corrosion, simplicity in processing, energy consumption, relatively low processing temperature, and recyclability. This has made PMCs the preferred choice of materials engineers for the development of products. PMCs consist of an engineering polymer as the matrix material. The matrix element is a continuous and homogeneous material into which the reinforcements are entrenched to amend the functioning of the matrix material. Commonly used matrix materials in PMCs are thermoplastics and thermosets.

### 9.1.1 Matrix Material

Thermoplastics melt and soften upon application of heat and are suitable for liquid flow forming. They derive their strength and stiffness from the monomer units, which are of very high molecular weight. Composites with thermoplastic matrices are generally manufactured by extrusion, followed by injection molding. Polyethylene (PE), polypropylene (PP), polystyrene (PS), polycarbonates (PCs), polyamides/nylons (PAs), polymethyl methacrylate (PMMA), polyacetal (POM), polyurethane (PU), polyvinyl chloride (PVC), polyimide (PI), polyetherimide (PEI), polyaryletherketones (PAEKs), polyamideimide (PAI), etc. come under this type.

*Synthesis and Tribological Applications of Hybrid Materials*, First Edition.
Edited by Mohammad Jawaid, Rajini Nagarajan, Jacob Sukumaran, and Patrick De Baets.

Thermoset matrices are high-density liquid polymers that are converted into hard brittle solids by polymerization, resulting in the formation of a covalently bonded three-dimensional network. The mechanical behavior of the composites made of thermosets depends on this networking. They do not melt on heating once curing is over. However, a loss in stiffness at the heat distortion temperature is noted. Phenol formaldehyde (PF), urea formaldehyde (UF), polyester, vinyl ester, and epoxy resins are some of the examples of thermoset polymeric materials. Of these, epoxy, vinyl ester, and unsaturated polyester resins cover a very broad class of thermosets. Compared to polyesters, epoxies have fair toughness and environmental resistance, low absorption of moisture, and considerably less shrinkage during curing [1].

### 9.1.2 Reinforcements

Many researchers have reported the synergistic performances of several PMCs reinforced with natural fibers such as bamboo, sugarcane bagasse, jute, kenaf, flax, grass, sisal, hemp, coir, ramie, and abaca, and synthetic fibers such as carbon, glass, aramid, nylon or rayon embedded in a polymer matrix and their performances have been evaluated by several researchers [2–7]. However, in order to enhance their performances two or more reinforcing elements, in the form of fibers/particulate fillers, are included in the polymer matrix, leading to the evolution of hybrid polymer matrix composites (HPMCs). The polymer matrix element surrounds and tightly binds the reinforcements, which are in the form of fibers or particulate fillers.

HPMCs represent a new class of alternative materials to conventionally filled, i.e. two-phase PMCs, which can overcome the problems associated with traditional composites [8, 9]. These HPMCs have been immensely used in thermal power sectors, sports, automotive, aircraft, military, and aerospace industries. They have exhibited significant weight savings for aircraft structures and remarkable resistance to corrosion and fatigue damage. Mercantile aircraft applications vary from minor flight control surfaces to more and more prime edifices [10]. Several investigators have studied the tribo performances of various HPMCs [11–17].

#### 9.1.2.1 Fiber Reinforcements

A fibrous reinforcement in HPMCs is categorized by its length being much larger than its cross section. The relation between the length and cross-sectional feature is known as aspect ratio. The diameter of fibers used in structural composites varies from 5 to 140 μm. The key tasks of the fibers are conveying the load, thermal stability, strength, and providing stiffness and other structural behavior. To achieve these tasks, the fiber in composite must possess increased ultimate strength, modulus of elasticity, and a little distinction of strength amid the fibers, increased stability of their strength during usage, and homogeneity of diameter. Carbon, glass, and Kevlar are the most important fibrous materials used for making composites. These fibers can be designed with unique architecture (fabrics) depending on the product requirement and manufacturing process. The fabrics are made of fibers orientated along two perpendicular directions. One is called the warp and the other is called the weft directions. The fibers are interlaced together, which means the weft fibers (tow) pass over and under the warp fibers, following a constant pattern. Intertwined fabric reinforced composites are becoming well known because of their well-adjusted behavior in the fabric plane as well as their ease of fingering during processing. Major types of weave forms of fabrics are illustrated in Figure 9.1.

***Plain Interweave Fabric*** Each warp fiber passes alternatively under and over each weft fiber as shown in Figure 9.1a. The fabric is symmetrical and possesses good stability. Nevertheless, it is most intricate to drape the interweaves, and the increased fiber gathering displays comparatively subdued mechanical behavior compared with the other styles of weaving.

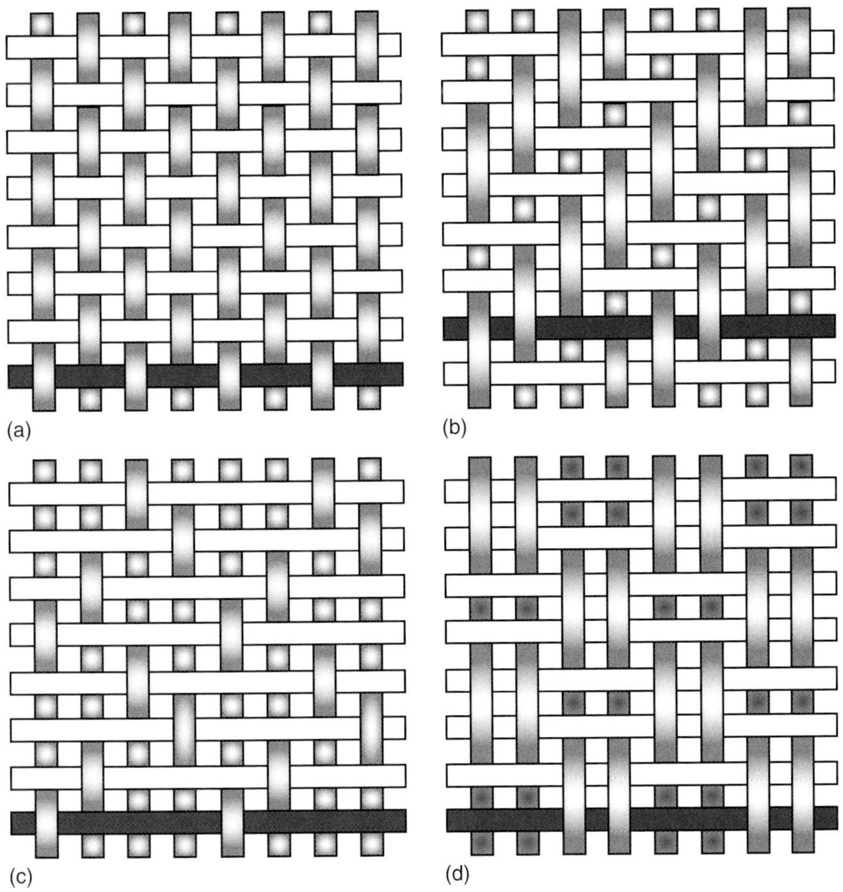

Figure 9.1 Different interweave forms of fabric: (a) plain, (b) twill, (c) satin, and (d) basket.

***Twill Interweave Fabric*** One or more warp fibers are sequentially interweaved over and under two or more weft fibers in a steady repetitive pattern as shown in Figure 9.1b. This is portrayed by excellent wet out and drape over the plain weave with a little decline in stability. With decreased fiber gathering, the fabric has a smoother surface with faintly better mechanical behavior.

***Satin Interweave Fabric*** They are basically twill interweaves customized to make a few junctions of warp and weft as shown in Figure 9.1c. It is designated by a harness number, which is the sum of fibers crisscrossed and conceded under, before the fiber recaps the pattern. Satin weaves are awfully flat, and have beneficial wet out and an improved level of drape. The decreased gathering of fibers in the interweave result in the demonstration of better mechanical behavior. Satin interweave fibers can be woven closely, which eventually produces fabrics with tight weave. But this weave style has low stability and is asymmetric in nature. The irregularity results in one face of the fabric having fibers running chiefly in the warp direction while the other face has fibers running chiefly in the weft direction.

***Basket Interweave Fabric*** They are similar to plain interweave except that two or more warp fibers consecutively intertwine with two or more weft fibers as shown in Figure 9.1d. This grouping of two warps crisscrossing two wefts is described as 2 × 2 baskets, but the grouping of fibers will essentially not be even. Basket interweave is relatively flat, and, through less crimp, sturdier than a plain interweave, but less steady.

#### 9.1.2.2 Particulate Reinforcements

Functional fillers of various shapes and sizes are used as reinforcing particles in HPMCs. The size ranges from a few microns to several hundred microns. Fillers in HPMCs can account for considerable weight saving. In comparison to resins and reinforcements, fillers are the least expensive of the major ingredients. A wide range of fillers are used starting from metallic powders to elastomeric fillers. Oxides such as aluminum oxide ($Al_2O_3$), zirconium dioxide ($ZrO_2$), silicon dioxide ($SiO_2$), copper oxide (CuO), copper sulfide (CuS), titanium dioxide ($TiO_2$), copper fluoride ($CuF_2$), lead sulfide (PbS), tungsten disulfide ($WS_2$), calcium sulfide (CaS), and boron nitride (BN) are some of the commonly used metallic fillers. Lubrication with liquids is limited in both technological and economic aspects such as physical and chemical degradation. Hence, solid lubricants such as graphite, molybdenum disulfide ($MoS_2$), lead sulfide (PbS), tungsten disulfide ($WS_2$), calcium sulfide (CaS), and polytetrafluoroethylene (PTFE) can be used.

Particulate fillers filled PMCs acquire better mechanical behaviors and these fillers are added from the techno-commercial aspects. The insertion of fillers in the polymer materials improves the physical, mechanical, and thermal behavior [18–22]. Even in composites reinforced with fiber, the addition of fillers has resulted in enhancement in behavior of the composites. For applications taking place in hard working conditions, such as friction bearings, chute liners used in power plants, gears, agricultural equipment, and pumps handling industrial fluids, the development of polymer based composites, which acquire high stiffness, toughness, and wear resistance, becomes crucial.

### 9.1.3 Friction and Wear

Wear is described as impairment to a solid surface, normally concerning continuing loss of material, due to relative motion with contacting surfaces. The major types of wear are adhesive, abrasive, fretting, erosion, and fatigue, which are generally witnessed in practical circumstances. Adhesive wear or dry sliding wear is caused by local adhesion between contacting solid surfaces, leading to material transfer between the two surfaces or loss from either surface. The friction and wear performances of PMCs are assessed by this wear test.

Abrasive wear is the most important among all forms of wear, because it contributes to about 63% of the total cost of wear [23]. Abrasive wear is caused by hard particles sliding on a softer solid surface and displaying or detaching material [24], in which hard asperities on one body moving across a softer body under some load penetrate and remove material from the surface of the softer body, leaving a groove. These hard asperities may in fact be discrete particles. Abrasive wear is classified as two-body abrasive wear (2-BAW) and three-body abrasive wear (3-BAW) according to the type of contact. It is straightforward to nominate a pair of terms such as constrained abrasive and free abrasive mode.

1) Two-body abrasion is where a hard rough body ploughs into a softer body as shown in Figure 9.2a.
2) Three-body abrasion wear is where a third body (usually hard granular matter) is placed between the sliding surfaces, which rolls and slides to cut grooves on the surface of the softer body as shown in Figure 9.2b.

Two-body wear is generally a low stress type of wear, where hard asperities plough into the soft surface. In three-body wear, due to the high stress, the hard abrasive particles between the two sliding surfaces roll and slide to result in ploughing, cutting, and cracking at the micro level as shown in Figure 9.3.

The wear mechanisms, i.e. micro-ploughing and micro-cutting, are chiefly ruled by plastic deformation of the wear material; however, micro-cracking mechanism results in fracture of the wear material. Therefore, the prevailing wear mechanisms that arise for a specific operational circumstance are persuaded to an excessive magnitude by the plastic deformation and fracture behavior of the wear material [25]. Three-body abrasion is commonly regarded as

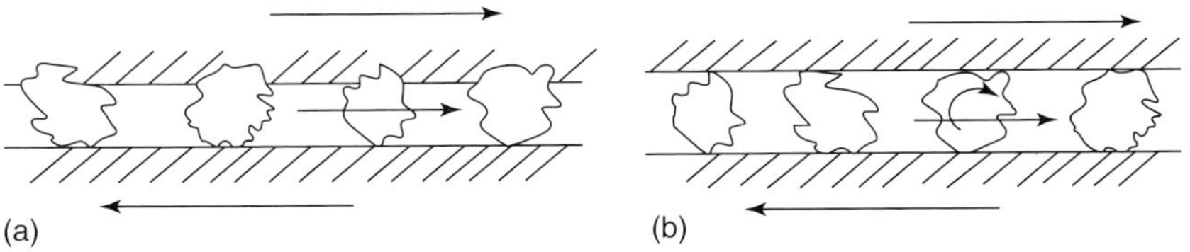

(a) (b)

**Figure 9.2** Schematic representations: (a) two-body abrasion and (b) three-body abrasion.

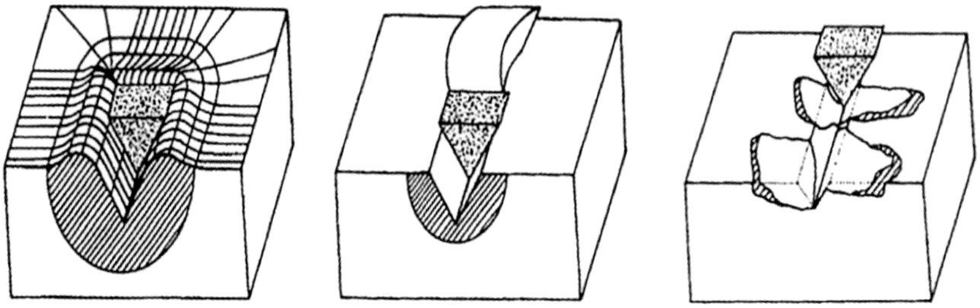


**Figure 9.3** Mechanisms involved in removal of material at microscopic level between material surface and abrasives: (a) micro-ploughing, (b) micro-cutting, and (c) micro-cracking.

more realistic: however, its study has attracted very less interest relatively. Investigations on abrasive wear performance of polymer composites are reported by a few researchers [12, 26–30]. However, the researchers have concluded that the abrasion process is a complex phenomenon as it is affected by various process parameters. Hence the authors have penned this chapter to share their acquired knowledge with the technical students, young budding researchers, and the dedicated research community on the tribological performance of hybrid carbon-epoxy PMCs with various reinforcing particulate fillers. Further, the wear mechanisms are discussed with the help of scanning electron micrograph images.

## 9.2 Materials and Methods

### 9.2.1 Matrix Material

In the present investigation, LY 556 epoxy resin (procured from Hindustan Ciba Geigy) is used as the matrix material, owing to its exceptional mechanical performance and good resistance to corrosion/chemicals. Another attribute, which has placed epoxies above others, is the easiest possibility for the addition of a curing agent, with/without the use of heat. LY 556 resin is a bifunctional epoxy resin, i.e. di-glycidyl ether of bisphenol-A (DGEBA). This was cured or hardened by the addition of HY 951, which is an aliphatic primary amine, viz. tri-ethylene tetramine (TETA) along with diluents DY 021, both purchased from Hindustan Ciba Geigy. They react with epoxide groups at room temperature and hence accomplish simple curing. The detailed compositions of the carbon-epoxy composite system are provided in Table 9.1.

### 9.2.2 Reinforcement Materials

The reinforcement material used in the present investigation is bidirectional carbon fiber fabric (Thornel™ T300). This fabric had a fiber area density of 208 $g\,m^{-2}$. The carbon fiber has high strength/tenacity with 3000 filaments per tow of the warp and weft of the fabric as shown in Figure 9.4. The yarns had a twist of

**Table 9.1** Constituents of matrix system.

| Constituent | Trade name[a)] | Chemical name[a)] | Epoxide equivalent[a)] | Density ($g\,cm^{-3}$)[a)] | Supplier | Parts by weight[a)] |
|---|---|---|---|---|---|---|
| Resin | LY-556 | Di-glycidyl ether of bisphenol A (DGEBA) | 182–192 | 1.16 | Hindustan Ciba Geigy Ltd. | 100 |
| Hardener | HY-951 | Tri-ethylene tetra amine (TETA) | — | 0.95 | Hindustan Ciba Geigy Ltd. | 10–12 |

a) From supplier's data sheet.

**Figure 9.4** Plain woven carbon fabrics.

15 turns per meter (TPM) and fineness (Tex) of 200 $g\,km^{-1}$. The diameter of the carbon fiber is 6–8 μm.

Bidirectional plain woven fabric is beneficial in load bearing in both longitudinal and lateral directions of the fabric plane. Woven reinforcement offers good stability in the warp and weft direction. The carbon atoms are bonded together in microscopic crystals and the crystal alignment makes the fiber very strong. Several thousand carbon fibers are twisted together to form a yarn, which may be used by itself or woven into a fabric as shown in Figure 9.4.

The different properties of the carbon fabrics (T300) used in the investigation are listed in Table 9.2. The properties exhibited by carbon fibers make them a suitable candidate for both structural and non-structural applications.

### 9.2.3 Particulate Fillers

The two particulate fillers used in the investigations are molybdenum disulfide ($MoS_2$) and aluminum oxide ($Al_2O_3$), where $MoS_2$ is regarded as lubricant while the $Al_2O_3$ is a ceramic filler.

Table 9.2 Properties of carbon fabric.

| Designation | Tensile strength (MPa)[a)] | Tensile modulus (GPa)[a)] | Elongation (%)[a)] | Density ($g\,cm^{-3}$)[a)] |
|---|---|---|---|---|
| T300 | 3530 | 230 | 1.5 | 1.76 |

a) From supplier's data sheet.

#### 9.2.3.1 Molybdenum Disulfide

Molybdenum disulfide ($MoS_2$) is a black solid, and is similar to graphite in appearance. It is slippery in nature or greasy to touch. It has a hexagonal layer-lattice structure. Molybdenum atoms are sandwiched between layers of sulfur atoms. Because of weak van der Waals interactions between the sheets of sulfide atoms, $MoS_2$ has a low coefficient of friction and is thus known as a solid or dry lubricant. $MoS_2$ powder of particle size 20–25 μm is used as particulate filler in the present study and is shown in Figure 9.5.

$MoS_2$ has advantages such as (i) practically no tendency to flow, creep, or migrate; (ii) minimum tendency to contaminate products or environment; (iii) awfully less volatility allowing it to be utilized in vacuum (e.g. space applications); (iv) chemical inertness allowing it to be utilized in reactive chemical situations; (v) normally steady to radioactivity (beneficial in nuclear power plants); (vi) fair load carrying ability; and (vii) nontoxicity. In contrast to graphite, $MoS_2$ does not depend on the existence of adsorbed fumes to act as lubricant.

#### 9.2.3.2 Aluminum Oxide

Aluminum oxide is artificially manufactured, and is a white or almost colorless crystalline material. It is also commonly used as fillers in different polymer matrices. Aluminum oxide ($Al_2O_3$) is absolutely insoluble in water. In its most usually occurring crystalline form, called corundum or $\alpha$-aluminum oxide, its hardness makes it fit for usage as an abrasive and as a constituent in cutting tools. Alumina is the most cost-effective and most widely used among the engineering ceramics. It has high strength and stiffness coupled with good hardness; its wear resistance to chemical attack enables it to be widely used as engineering ceramics. A combination of excellent properties and low cost has made fine grade alumina to be used in a wide range of applications. $Al_2O_3$ powder of particle size 20–25 μm is used as particulate filler in the present study and is shown in Figure 9.6.

Figure 9.5 Molybdenum disulfide particulate filler.

**Figure 9.6** Aluminum oxide particulate filler.

The important properties of alumina are (i) good hardness and wear resistance, (ii) excellent dielectric properties, (iii) good thermal conductivity, (iv) high strength and stiffness, (v) resistance to strong acid and alkali attack at high temperatures, and (vi) excellent size and shape capability. These key properties of alumina found varied applications in different fields such as wear pads, seal rings, abrasion resistant tube and elbow liners, gas laser tubes, high-temperature and high-voltage insulators, furnace liner tubes, grinding media, and many more. All these fillers are treated with a 2% organo-reactive silane coupling agent. Table 9.3 presents the physical properties of the fillers used in the present study.

### 9.2.4 Composite Fabrication

The composite materials deliberated in the current discussion comprise of bidirectional carbon fabric of around 6–8 μm diameter as reinforcement. LY 556 epoxy resin with HY 951 grade room temperature curing hardener with diluents DY 021 mixture was cast off for the matrix material. Predetermined quantity of epoxy resin and the fillers were assorted by means of high shear mixture (T-T18 ULTRATRURRAX Basic) at an operative speed of 2000 rpm for 10 min. A temperature of about 50 °C was maintained during mixing. To obtain 3-mm-thick laminates eight layers of fabrics were used. The carbon fabric reinforced epoxy (C-E) composites were prepared by hand lay-up technique followed by the autoclave molding process (pressure 0.75 MPa and temperature 90 °C). Autoclave curing is preferred over other curing, to prevent voids and to obtain good surface

**Table 9.3** Physical properties of the fillers.

| Filler | Color[a] | Size (μm)[a] | Density ($g\,cm^{-3}$)[a] |
|---|---|---|---|
| Molybdenum disulfide | Black | 20–25 | 5.06 |
| Aluminum oxide | White | 20–25 | 3.69 |

a) From supplier's data sheet.

finish. Figure 9.7 shows the different stages involved in the fabrication of the composite under investigation. Table 9.4 illustrates the particulars with respect to designation and weight percentage of the carbon fabric, epoxy, and micro-fillers utilized in this study.

### 9.2.5 Dry Sliding Wear Test

The pin-on-disk apparatus was used for the characterization of sliding wear by evaluating friction and wear. The test device designed and developed by Magnum Engineers, Bangalore, India, is mainly envisioned for defining the tribological features of an extensive choice of materials under circumstances of varied normal loads and velocities, and is depicted in Figure 9.8. A static pin fixed on a pin holder is brought into interaction beside a rotating disk at an indicated speed as the pin slides, causing frictional force acting between the pin and disk. A graphic

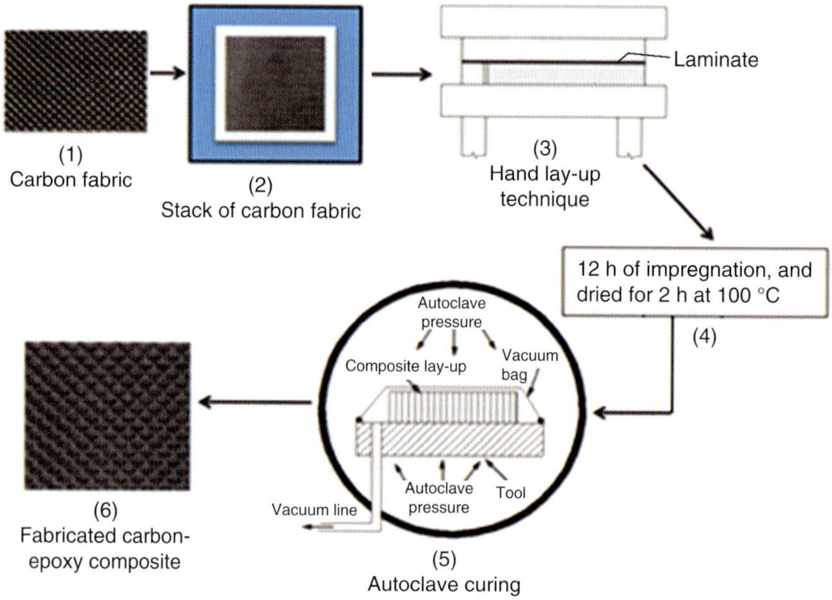


**Figure 9.7** Stages in composite fabrication.

**Table 9.4** C-E composites utilized in the current study.

| Material description | Designation | Carbon fabric (wt%) | Matrix (wt%) | Filler (wt%) |
|---|---|---|---|---|
| Carbon-epoxy | C-E | 60 | 40 | — |
| 5% $MoS_2$ filled C-E | C-E-5M | 60 | 35 | 05 |
| 10% $MoS_2$ filled C-E | CE-10M | 60 | 30 | 10 |
| 5% $Al_2O_3$ filled C-E | CE-5A | 60 | 35 | 05 |
| 10% $Al_2O_3$ filled C-E | CE-10A | 60 | 30 | 10 |

**Figure 9.8** Pin-on-disk apparatus.

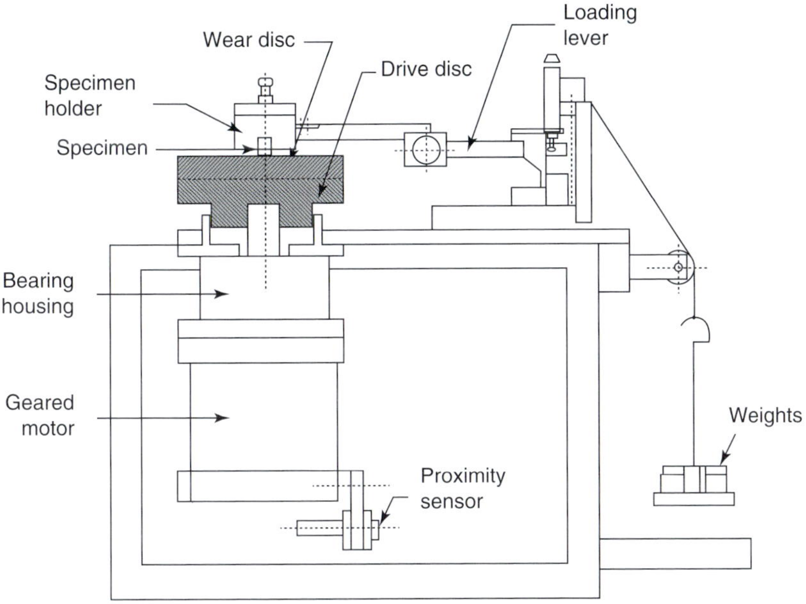


**Figure 9.9** Graphic representation of pin-on-disk test device depicting the various elements.

representation of the pin-on-disk test rig, presenting the various elements, is shown in Figure 9.9. The applied normal load, wear track diameter, and rotational speed are followed in accordance with ASTM G99-17 [31] test standard.

The surface of 5 mm × 5 mm × 3 mm composite sample, attached to a mild steel pin of 6 mm diameter and 25 mm length called as pin assembly, interacts with the disk made of EN32 steel of hardness 65 HRC having dimensions 160 mm diameter, 8 mm thickness, and 0.1–0.2 μm surface roughness ($R_a$).

Prior to experimentation, the test specimens were polished against 600 grade SiC paper to confirm apt interaction with the counterface. The surfaces of both the specimen and the disk were cleansed with a soft paper wetted in acetone and completely dried before the test. The pin assembly was primarily and after the slide wear test, it was weighed in an electronic balance (Mettler Toledo) of accuracy of 0.1 mg. Three specimens were run for a particular set of conditions and the average of the three readings was taken for further analysis. The wear loss measured was then converted to wear volume utilizing the measured density of the sample. The details of dry sliding wear test factors utilized in the current study are enumerated in Table 9.5.

The following equations were used to calculate wear volume loss and specific wear rate.

$$\text{Wear volume loss} = \Delta V = \frac{\text{Mass loss}}{\text{Density}} = \frac{m_1 - m_2}{\rho} \tag{9.1}$$

$$k_s = \frac{\Delta V}{L \times D} \tag{9.2}$$

where $m_1$ is the mass of the specimen before the test, $m_2$ is the mass of the specimen after the wear test, $\rho$ is the density of the composite in g mm$^{-3}$, $\Delta V$ is the volume loss in mm$^3$, $L$ is applied load in Newton, and $D$ is the sliding distance in meter.

### 9.2.6 Three-Body Abrasive Wear Test

Three-body abrasion experiments were carried out using a dry sand/rubber wheel abrasion tester (RWAT) in accordance with ASTM G65-17 [32] standard except for modification to the applied normal load as shown in Figure 9.10.

The schematic diagram of the same is portrayed in Figure 9.11. The dry sand particles of AFS 60 grade were used as abrasives and they are jagged in shape with sharp edges. The standard dimensions (in millimeters) of the composite test specimen used in three-body abrasion are depicted in Figure 9.12. The abrasive particles were forced between the test specimen and the rotating rubber wheel (rotational speed 200 rpm), made of chlorobutyl material with a hardness of 58–62 Shore A. The feed rate of the abrasive particles was in the range of 255–265 g min$^{-1}$. The specimen was cleaned with acetone in an ultrasonic cleaner and then dried. Its initial weight was measured in a digital measuring balance – Mettler, Toledo – with an accuracy of 0.1 mg and the specimen was mounted in the holder. After the test, the specimen was removed, thoroughly cleaned, and again the final weight was measured. The difference in

**Table 9.5** Particulars of the slide wear test features employed in the current study.

| Test conditions | Units | Values |
|---|---|---|
| Sliding velocity | m s$^{-1}$ | 0.5, 1.0, 1.5, and 2.0 |
| Normal load | N | 20, 40, 60, and 80 |
| Sliding distance | m | 3000 |

**Figure 9.10** Dry sand/rubber wheel abrasion tester.

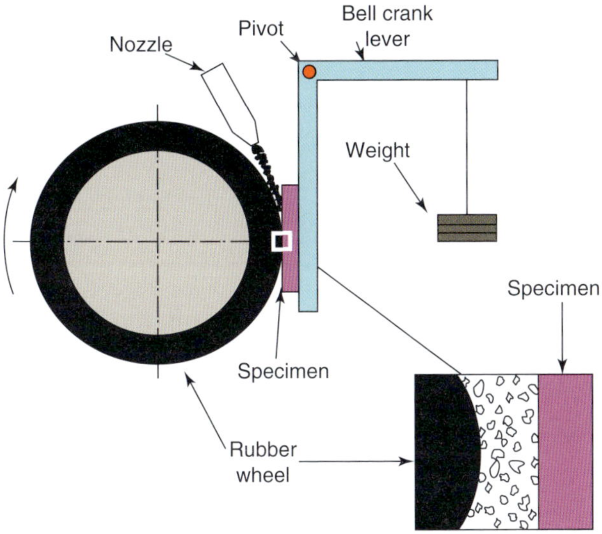


**Figure 9.11** Schematic diagram of the three-body abrasive test setup.

weight calculated is a measure of the abrasive wear loss. At least three tests were performed and the average value obtained was used in the analysis.

The particulars of 3-BAW test factors utilized in the current discussion are enumerated in Table 9.6. The investigations were conducted at two different loads (23 and 34 N) at an unceasing sliding velocity of 2.15 $m\,s^{-1}$. Further, the abrading distances were varied in steps of 270 m, from 270 to 1080 m. The abrasive wear was determined by the weight loss, and then transformed into wear volume utilizing the assessed density data. After the wear test, the sample was again cleaned. The wear volume loss ($\Delta V$) and specific wear rate ($K_s$) were calculated using Equations (9.1) and (9.2) respectively.

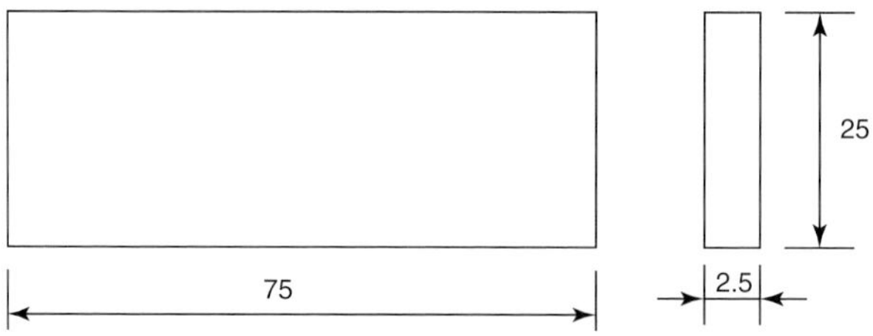


**Figure 9.12** Three-body abrasion test specimen (all dimensions are in millimeter).

**Table 9.6** Particulars of 3-BAW test in the current discussion.

| Test conditions | Units | Values |
|---|---|---|
| Load | N | 23, 34 |
| Abrading distance | m | 270, 540, 810, 1080 |
| Speed | $m\,s^{-1}$ | 2.15 |

## 9.3 Results and Discussion

The present work attempts to produce and evaluate the characteristics of particulate filled and C-E composites for tribological appliances. Earlier research reviews on wear of polymeric composites demonstrate anisotropic characteristics. Besides, investigations on the effect of micro-fillers loading and applied normal load/abrading distance on wear rate revealed some discrepancies. Therefore, the friction and wear performance of unfilled C-E and C-E composites filled with $MoS_2$ and $Al_2O_3$ have been reviewed in terms of the coefficient of friction ($\mu$), wear volume loss, and specific wear rate ($K_s$).

### 9.3.1 Dry Sliding Wear Performance of Carbon-Epoxy Composites

#### 9.3.1.1 Wear Volume Loss

The plot of wear volume loss of unfilled and particulate filled C-E composites as a function of applied normal load is illustrated in Figure 9.13a–d. The wear volume loss increases with increase in load and velocity for all the composites. The wear volume loss as a function of velocity for unfilled and filled C-E composites is shown in Figure 9.14a–d.

Wear volume loss increases with increase in sliding velocity for all the composites and decreases with increase in filler loading, as depicted in Figure 9.15, which highlights the beneficial effect of inclusion of micro-fillers into the C-E composite. The wear volume loss of unfilled C-E composite is the maximum among all the composites examined. It was also observed that inclusion of particulate micro-fillers to the C-E composite reduced the wear volume loss. The wear volume decreased with increase in filler loading, and 10 wt% $MoS_2$ filled C-E (C-E-10M) composite showed the least wear volume loss under different loads. The reduction in wear volume loss due to the addition of particulate fillers such

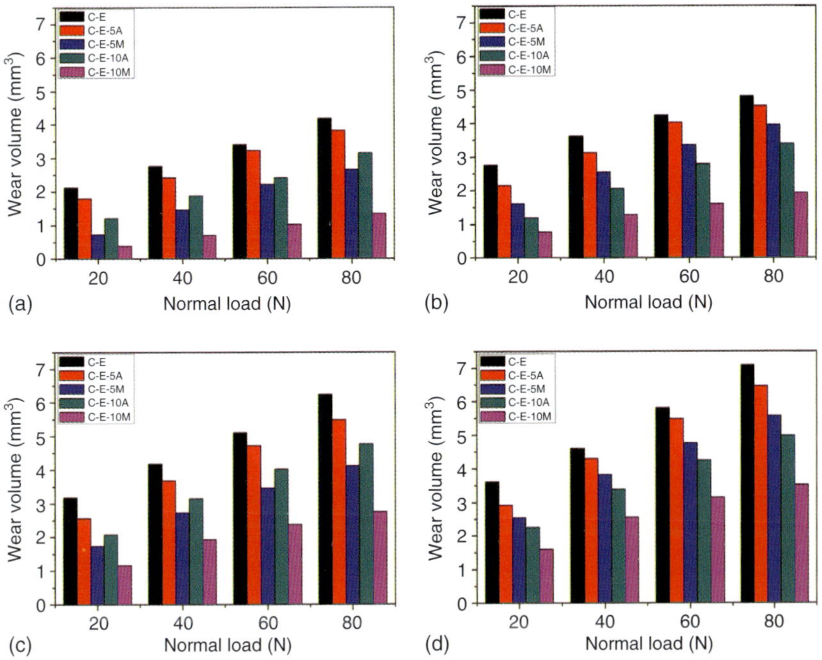


**Figure 9.13** Wear volume loss of unfilled and particulate filled C-E composites as a function of load and for different velocities: (a) 0.5 m $s^{-1}$, (b) 1 m $s^{-1}$, (c) 1.5 m $s^{-1}$, and (d) 2 m $s^{-1}$.

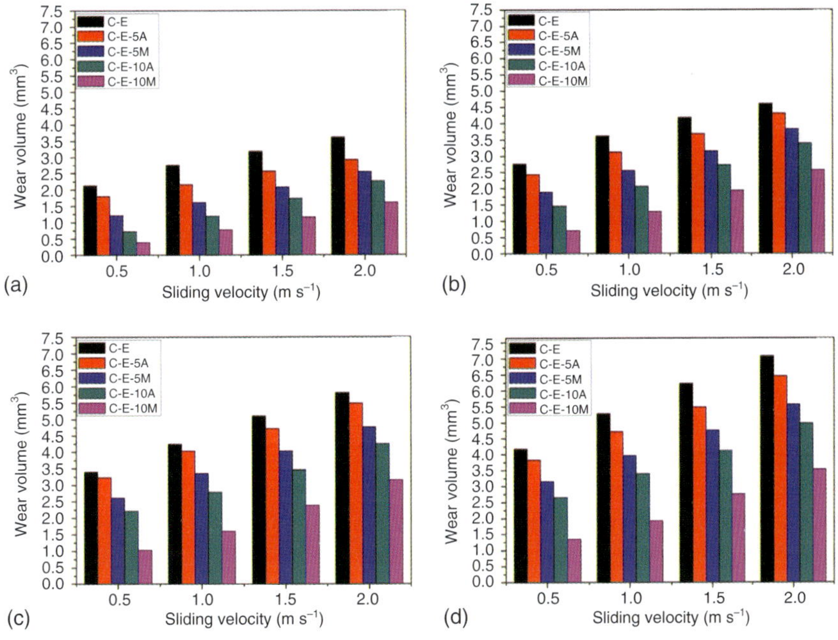


**Figure 9.14** Wear volume loss of unfilled and particulate filled C-E composites as a function of sliding velocity and for different loads: (a) 20 N, (b) 40 N, (c) 60 N, and (d) 80 N.

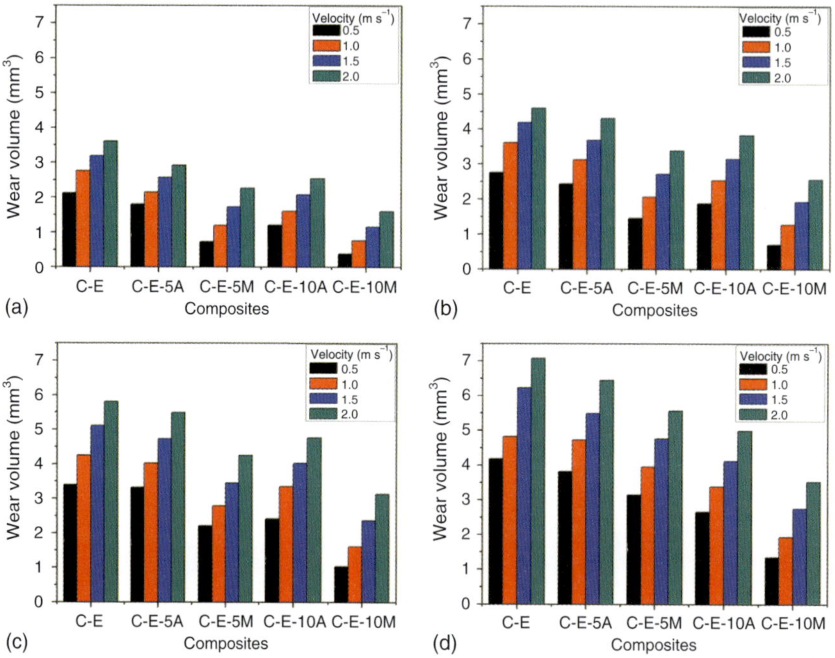


**Figure 9.15** Wear volume loss of unfilled and particulate filled C-E composites for different velocities and loads: (a) 20 N, (b) 40 N, (c) 60 N, and (d) 80 N.

as $Al_2O_3$ and $MoS_2$ can be accredited to the existence of particulate fillers that act as active blockades to avoid significant scale fragmentation of the epoxy matrix. It can be realized from Figure 9.13a–d that the wear volume loss of composites with 10 wt% $MoS_2$ (C-E-10M) decreased about by 68% compared to unfilled C-E under a sliding velocity of 0.5 m s$^{-1}$ and at 80 N load. In the case of C-E with 10 wt% $Al_2O_3$ (C-E-10A), the wear volume loss was approximately reduced by about 36% that of unfilled C-E. It must be observed that the decrease in wear volume loss was superior in $MoS_2$ filled C-E composites. The wear volume loss for C-E-10M was 1.34 mm$^3$ under a sliding velocity of 0.5 m s$^{-1}$ while that of C-E-10A was 2.66 mm$^3$. In addition, it can be seen that the wear volume loss of unfilled and particulate filled C-E composites increased with increase in sliding velocities from 0.5 to 2 m s$^{-1}$. Unfilled C-E and C-E-5A composites showed excessive wear volume loss after 1.5 and 2 m s$^{-1}$ in the investigation, and the corresponding wear fragments were in the form of large wedges under a load of 80 N and at 2 m s$^{-1}$ sliding velocity (Figure 9.13). These results are further substantiated with scanning electron microscopy (SEM) micrographs in Section 9.3.1.4.

It is known that in polymer composites wear resistance can be essentially increased by decreasing the coefficient of friction or by increasing their hardness. In the present work, the outcomes exhibit that the wear resistance of unfilled C-E was increased by addition of $Al_2O_3$ and $MoS_2$ microparticles. In the study of sliding wear behavior of epoxy matrix composites filled with silica particles by Xing and Li [33], the neat epoxy behaved in a brittle manner and macrocracks were formed, resulting in larger wear debris. However, with the addition of micro-silica particles into the epoxy matrix, the propagation of cracks was prevented by the particles near the surface layer. Hard filler particles have the ability to partially support normal loads during sliding, which eventually reduces the wear volume of the PMC. Similar observations were also made in the dry sliding wear tests on PTFE filled $MoS_2$/graphite composites [34].

Molybdenum disulfide ($MoS_2$) is a dark blue–gray or black solid, and has a hexagonal layer-lattice structure [35, 36]. The layer comprises flat sheets of atom or molecules, and the configuration is termed a layer-lattice configuration. The significant outcome is that the materials can shear effortlessly parallel to the layers than crosswise. They can thus assist fairly hefty loads at right angle to the layers while still being able to slide easily parallel to the layers. This characteristic is being successfully applied for lubrication activity. Unlike graphite, $MoS_2$ does not depend on the existence of adsorbed vapors to act as lubricant. Hence, it can be used adequately at elevated vacuum and temperatures. $MoS_2$ starts to oxidize at 350 °C in air, although it can even be employed at small intervals up to 450 °C. The oxidation generates molybdic oxide ($MoO_3$), which is a good lubricant at elevated temperatures but wears quickly. Apart from oxidation, it is stable with most chemicals; however, it is affected by powerful oxidizing acids and alkalis. To sum up, $MoS_2$ is a versatile and beneficial material where oils or greases cannot be expended or do not have adequate load-carrying capability [37]. Li et al. [38] studied the wear diminution mechanism of graphite and $MoS_2$ in epoxy composites. They found that $MoS_2$ with lamellar structure becomes $MoO_3$ during the wear process because of the thermal influence on friction and stresses of compression or shear, and hence $MoS_2$ has a minute influence on the coefficient of friction. Finally, they concluded that the presence of $MoS_2$ can confine the transportation of the matrix to the steel counterface, so that the wear resistance of the composite increases effectively.

#### 9.3.1.2 Specific Wear Rate

The specific wear rates ($K_s$) of unoccupied and particulate occupied C-E composites with respect to normal load and for various sliding velocities are illustrated in Figures 9.16 and 9.17 respectively. The $k_s$ of all the composites decreases with increase in load and increases with increase in sliding velocity. The $k_s$ of unfilled C-E was the highest for all loads/sliding velocities. It was also noticed that the increase in filler loading led to decrease in $k_s$ for constant load and velocity. The 10 wt% $MoS_2$ occupied C-E (C-E-10M) demonstrated least $k_s$ over $Al_2O_3$ occupied and unoccupied C-E composites. The addition of particulate fillers into C-E composite reduces the wear volume and $k_s$. The reduction in $k_s$ was due to the formation of a transfer film on the counterface. Kishore and Kumar [39] noticed that the inclusion of alumina particles (size <1 μm) into epoxy increased the wear resistance. Durand et al. [40] investigated the sliding wear performance of epoxy matrix composites reinforced with several kinds of micron-sized ceramic particles (size ranging from 5 to 100 μm), such as TiC, $Al_2O_3$, TiN, SiC, $ZrO_2$, and $TiO_2$. Pettarin et al. [41] studied the sliding wear behavior of $MoS_2$ filled high molecular weight, high-density polyethylene matrix. They reported that the wear behavior was altered by the addition of $MoS_2$ and it was also reported that the composites demonstrated the lowest wear rate for 10 wt% of

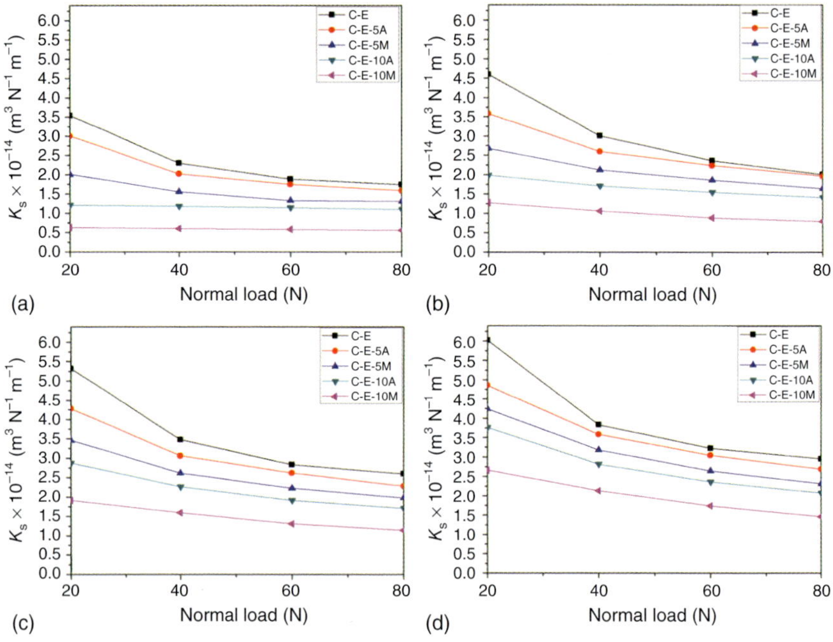


**Figure 9.16** Specific wear rate of unoccupied and particulate occupied C-E with respect to load and for different velocities: (a) 0.5 m s$^{-1}$, (b) 1 m s$^{-1}$, (c) 1.5 m s$^{-1}$, and (d) 2 m s$^{-1}$.

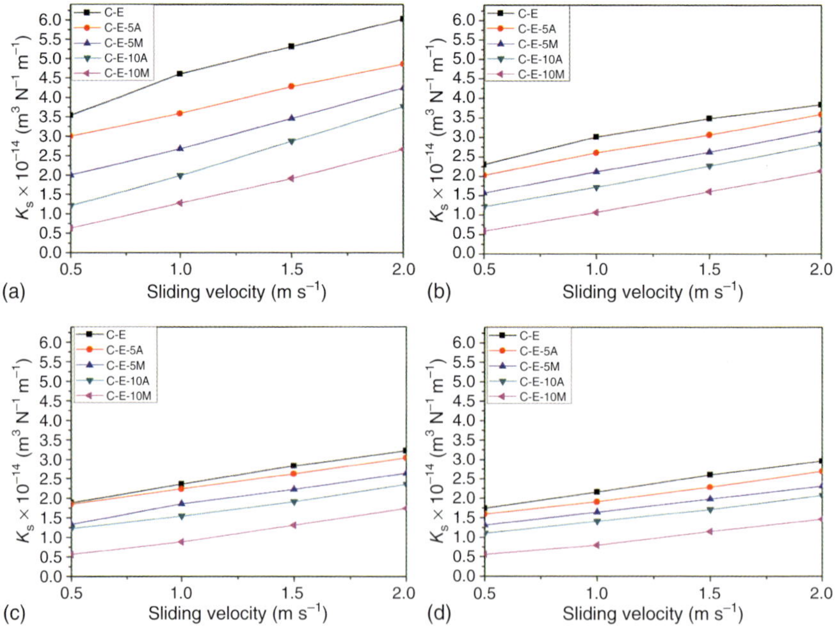


**Figure 9.17** Specific wear rate of unoccupied and particulate filled C-E with respect to velocity, and for different loads: (a) 20 N, (b) 40 N, (c) 60 N, and (d) 80 N.

$MoS_2$ and the same is accredited to the development of a uniform and adherent transfer film on the counterface. The wear diminution mechanism of graphite and $MoS_2$ fillers in epoxy composites was investigated by Li et al. [38] and they reported that the graphite in epoxy matrix can reduce the coefficient of friction effectively and increase the wear resistance of the composite. In $MoS_2$ filled epoxy composites, the coefficient of friction cannot be decreased but the wear resistance can be efficiently increased by controlling the epoxy transfer from the surface of the composite to the steel counterface.

The dry sliding wear behavior of nylon 1010 reinforced with short carbon fiber and filled with $MoS_2$ was studied by Wang et al. [42]. They revealed that the accumulation of carbon fiber was effective in decreasing friction and wear, but the $MoS_2$ filler increased its wear. Wear and friction diminution was more substantial while the carbon fiber was used as reinforcement along with $MoS_2$ filler. The ability of the synergistic fillers in facilitating the development of a thin, identical, and continuous transfer film would help improve the wear resistance. The influence of fillers such as graphite and silicon carbide particles (SiC) on the dry sliding wear of glass fabric reinforced epoxy (G-E) composites was investigated by Suresha et al. [43], and the study revealed that wear loss increases with increase in sliding velocity for both filled and unfilled G-E composites.

The magnitude of wear loss was less in filled G-E than in unoccupied G-E composites. The SiC filled G-E composite exhibited maximum resistance to wear. The increased exposure of the reinforcing fibers to the counterface resulted in increased fiber fracture. Graphite filled G-E composite exhibited reduced coefficient of friction and this can be accredited to the existence of graphite – a solid lubricant. In the recent studies of dry sliding wear of C-E filled with graphite, Suresha et al. [44] demonstrated that specific wear rate increased with increase in sliding velocity and the specific wear rate of graphite filled C-E was lower than that of the unoccupied C-E composite. The decrease in the wear rate of graphite filled C-E composites was due to the growth and expansion of the transfer film.

In the present work, the wear volume loss data indicate that $MoS_2$ filler could improve the transfer film of filled C-E to steel counterface. For unoccupied and $Al_2O_3$ filled C-E composites, the transfer film was relatively thin, non-uniform, and uninterrupted on the counterface surface. The cause for this could be the quantity of fiber and hard $Al_2O_3$ particles in epoxy, which grazed the transfer film partly. When 10 wt% $MoS_2$ was included to the composite with 60 wt% carbon fiber, the creation of transfer film was thin, even, and uninterrupted. From the SEM micrographs (Section 9.3.1.4), it can be determined that the synergism between carbon fiber and $MoS_2$ filler may help to form a good and persistent transfer film on the counterface surface, thus enhancing the resistance to wear and decreasing the friction coefficient.

#### 9.3.1.3 Coefficient of Friction

The graph of friction coefficient ($\mu$) with respect to normal load for filled and unoccupied C-E composites is presented in Figure 9.18. For unoccupied C-E, the $\mu$ for lower loads remained almost constant, and at higher loads it increased

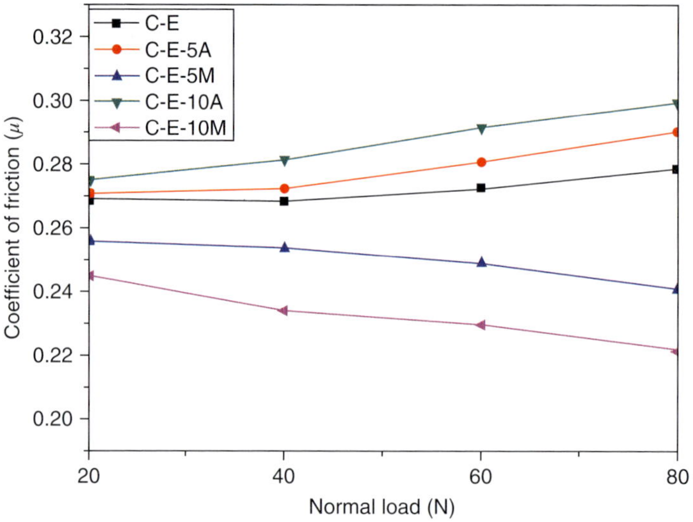


**Figure 9.18** Friction coefficient with respect to normal load for unoccupied and particulate filled C-E composites.

marginally. The $\mu$ for $Al_2O_3$ filled C-E composites increased and the increment was high for C-E-10A sample, while $\mu$ of $MoS_2$ filled C-E composites decreased and the maximum reduction was obtained for the C-E-10M sample.

The wear loss and $\mu$ of particulate filled and unfilled C-E composites are closely linked to the formation and development of a transfer film on the counterface. During sliding of the unfilled C-E composite, the epoxy wears and only bidirectional carbon fibers are in interaction with the counterface. As normal load increases, the interface temperature increases, leading the carbon fiber to fracture, resulting in a complex condition at the interface. The fractured and debonded fibers act as the third body [45] resulting in increased abrasion. This leads to marginal increment in $\mu$ in the unfilled C-E composite. In the case of $Al_2O_3$ filled C-E composite, when epoxy comes in contact, adhesive wear occurs and further sliding leads to the detachment of particulate filler along with carbon fibers. The interface during the later part of dry sliding consists of $Al_2O_3$ particles and carbon fibers. The presence of hard ceramic particles in the interface of the sample and counterface increases the $\mu$. But for $MoS_2$ filled C-E composites, the interface is filled with solid lubricant $MoS_2$ particles and during further sliding these particles smear the interface, forming a transfer film that reduces the $\mu$. A similar trend in the dry sliding of nylon 1010 reinforced with $MoS_2$ and short carbon fiber was observed by Wang et al. [42]. They conducted tribo-chemical studies by X-ray photoelectron spectroscopy (XPS), which revealed that $MoS_2$ decomposed, and $MoO_3$, FeS, $FeSO_4$, and $Fe_2(SO_4)_3$ were generated through sliding. It was determined that FeS, $FeSO_4$, and $Fe_2(SO_4)_3$ compounds could enhance the adhesion between the transfer film and the counterface surface. The formation of a thin, uniform, and continuous transfer film would be a factor contributing to the decrease in $\mu$ of nylon composites.

#### 9.3.1.4 Worn Surface Morphology

The SEM examination of worn surfaces of unoccupied and particulate filled C-E composite specimens against steel counterface under a sliding velocity of 0.5 m s$^{-1}$ and normal load of 20 N is shown in Figure 9.19a. The filled arrow marks show the sliding direction in Figures 9.19 and 9.20.

The matrix debris generation and fiber matrix debonding start at different locations. The worn surface is relatively smooth and some of the exposed fibers are worn. The fiber micro-cracks and fiber breakage are very less. When the sliding velocity and applied normal loads are high at 2 m s$^{-1}$ and 80 N respectively, the worn surface feature shows additional fiber breaking and removal of fibers from the surface along with increased matrix debris accumulation (Figure 9.19b). At higher sliding velocity and normal loads, the contact temperature increases, which results in the rupture of the matrix, particularly at the interfacial region. This results in increased surface damage with ripping of the matrix and wreckage formation. Hence, the wear volume loss of unoccupied C-E composite is high and it enhances with increase in sliding velocity and applied normal load. The different stages in the wear of fiber reinforced polymer composites as stated by Tsukizoe and Ohmae [45] can be observed such as wear thinning of fibers (marked A), breakdown of fibers (marked B), and peeling off of the fibers from the matrix (marked C).

Figure 9.20a,b presents the worn surfaces of $MoS_2$ particulate occupied C-E composites with different loads and sliding velocities. The 10 wt% $MoS_2$ filled C-E composite sample subjected to low sliding velocity (0.5 m s$^{-1}$) and low applied normal load (20 N) is shown in Figure 9.20a. Small distinct parallel grooves (marked J) formed due to the ploughing of asperities present on the disk along with matrix breakage (marked K) at different places are seen. The increased applied normal load and sliding velocity effect on C-E-10M are shown in Figure 9.20b. It shows debonding and peeling off of the fiber from the matrix material and fiber rupture. These features observed when compared with the unfilled C-E sample at same operating conditions showed less damage. Particulate fillers used in the composite were silane treated, which improves the interfacial adhesion of matrix. Additionally, the particulate filler $MoS_2$ acts as a solid lubricant.

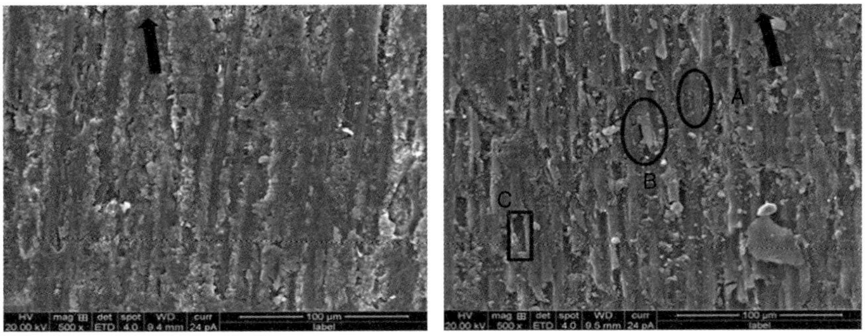


(a) (b)

**Figure 9.19** SEM of the worn surfaces of unoccupied C-E samples at (a) 0.5 m s$^{-1}$, 20 N and (b) 2 m s$^{-1}$, 80 N.

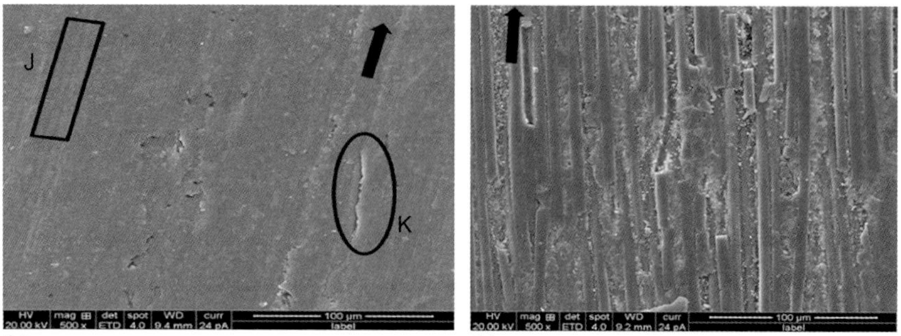

(a) (b)

**Figure 9.20** SEM of the worn sample of $MoS_2$ filled C-E samples at (a) $0.5\,m\,s^{-1}$, 20 N and (b) $2\,m\,s^{-1}$, 80 N.

## 9.3.2 Abrasive Wear Performance

### 9.3.2.1 Abrasive Wear Volume Loss

Figure 9.21a,b presents the wear volume loss with abrading distance for diverse loads. It is obvious from these figures that regardless of the kind of filler material utilized, there is a near linear trend of wear volume loss. The filled (both $MoS_2$ and $Al_2O_3$) C-E composites demonstrated significantly inferior wear volume loss than the unoccupied C-E. Furthest, the wear volume loss of $MoS_2$ filled C-E is much less than that of $Al_2O_3$ filled and unoccupied C-E composites.

Diverse kinds and intensity of fiber reinforcement and/or solid lubricants are used to increase several physical and/or mechanical and tribological behavior of polymers. Harsha et al. [29] studied 2-BAW and 3-BAW of PAEK composites. They reported that fillers such as PTFE and graphite were unfavorable to abrasive wear resistance related to glass fiber PAEK composites. Suresha et al. [28, 30] studied the 3-BAW of graphite/SiC filled G-E composites and determined that the graphite filler increased the specific wear rate and SiC reduced the specific wear rate of glass–epoxy composite. Suresha et al. [43] also studied the dispersed silane treated graphite particles in C-E composites. The resulting composites showed improved mechanical properties and improved abrasion resistance. Bijwe and coworkers [46] demonstrated that alumina is quite effective in lowering the wear rate of phenolic resin. The effect of fibers and/or fillers on the abrasive wear resistance of pure polymers was intricate and variable and diverse trends were registered by researchers. Lancaster [47] investigated 13 polymers reinforced with 30 wt% short carbon fiber and stated that the wear behavior of 7 composites improved with reinforcement and that of 6 composites reduced. The prime motive for including the fillers is to enhance the mechanical behavior, but the influence on tribo-behavior is not always beneficial. The tribo-mechanical performance of polymer reinforced composites depends upon the type of filler, geometry, and chemical treatment. The role of inclusion of molybdenum disulfide on tribological behavior of high molecular weight, high-density polythene was studied by Pettarin et al. [41]. They adopted dry sliding and abrasive wear tests to determine the wear performance of HMWHDPE composites. Adhesive wear mechanism was dominant and characterized by the formation of

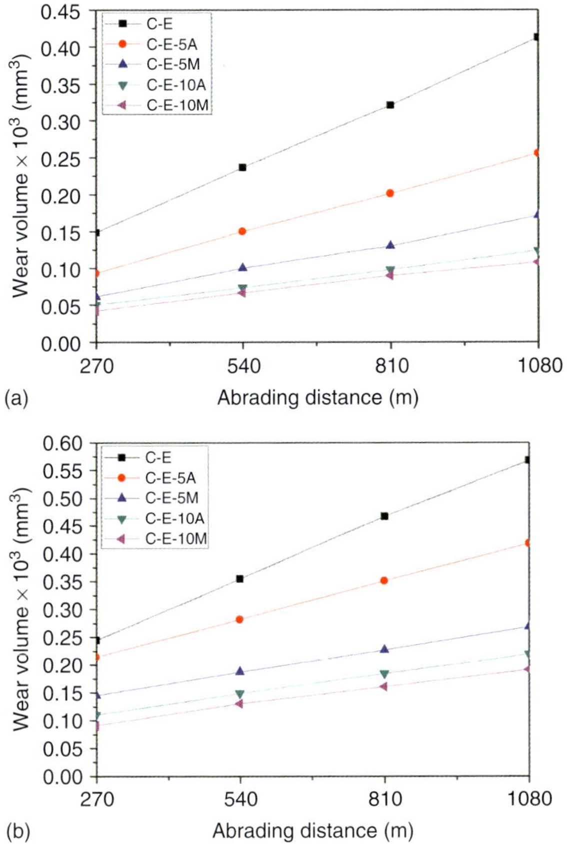


**Figure 9.21** Wear volume versus abrading distance of particulate filled C-E composites at (a) 23 N and (b) 34 N.

an even and adherent transfer film in dry sliding wear test. However, under abrasive situations an encouraging rolling effect of molybdenum disulfide particles was observed. In the present work, based on earlier literature, the $MoS_2$ and $Al_2O_3$ fillers were altered with organo-reactive silane treatment, which is certain to improve the interfacial bonding with the polymer and fiber. The $MoS_2$ modified by silane treatment filled in C-E composite demonstrated significantly less wear than $Al_2O_3$ filled and unoccupied C-E composites. This is because the carbon fiber reinforcement in epoxy reduces the abrasive wear resistance owing to debonding and ripping at the carbon/matrix interface. Further, the wear volume damage of $MoS_2$ filled C-E composite is much less than that of unoccupied C-E composite. The fibers (warp fibers) in C-E composites that were aligned along the abrading direction could have worked as a barricade in fracturing the crosswise fibers (weft fibers) as they were intertwined with the warp fibers.

Surface treatments of micro-fillers influence the tribological properties of polymer composites. The present write up focuses on the 3-BAW behavior of the composites consisting of epoxy, carbon fiber, and $Al_2O_3/MoS_2$ microparticles with silane treatment. Aluminum oxide is renowned for its hardness and is frequently utilized as a grinding agent. As pondered by Schwartz and Bahadur [48], it would not be an apt filler in the micro-scale powder structure, owing to its jaggedness, which tends to scratch the mating counterface. The material in the submicron or nanoscale powder structure possesses considerably less jaggedness and hence is not that abrasive. Kishore and Kumar [39] noticed that the inclusion of $Al_2O_3$ particles (size <1 μm) into epoxy escalates the sliding wear resistance of the material. Our earlier work indicated that glass fiber reinforced epoxy resin (G-E) with unmodified $Al_2O_3$ (fine particles) exhibits higher efficiency than graphite in advancing the abrasion resistance of the G-E composite [30]. The abrasive wear properties of G-E composites filled with $Al_2O_3$, SiC, and pine bark dust were examined by Patnaik et al. [49]. They have determined that the wear rate is susceptible to the type of filler used. The results showed that the G-E composite filled with SiC filler demonstrated the lowest wear resistance and the fibers were detached effortlessly with least resistance to the action of abrasive particles, whereas G-E filled with pine bark dust filler showed highest resistance to the material elimination related to other composites used.

Molybdenum disulfide is a well-known solid lubricant [50]. Its lubrication capability, i.e. capacity for effortless cleavage with little friction, is innate to its crystal layered edifice [51]. Each crystal layer comprises two coatings of sulfur atoms detached by a film of molybdenum atoms. The atoms resting on a similar crystal film are closely stored and intensely joined to each other, and the coatings themselves are separated comparatively far off. For example, the van der Waals bonds are weak. Certain manufacturers of $MoS_2$ modified polymers attribute the increase in wear resistance to the nucleating effect of the filler [52] and not to its lubrication effect. In the present study, the wear volume was low in the $MoS_2$ filled C-E composite and it can be accredited to the intrinsic mechanical performance and self-lubricating characteristic of $MoS_2$. Further, the interaction between the surface treated $MoS_2$ units and the epoxy leads to improved adhesion because of better polymer–filler activities. The effect of fibers and/or fillers on the abrasive wear resistance of unmixed polymer was considerably complex, and erratic and diverse developments were observed [47, 53]. Thus, in the present work microparticles of $Al_2O_3/MoS_2$ into C-E showed different wear rates. The $Al_2O_3$ filled C-E composite demonstrated poor resistance to abrasion owing to its jaggedness, which tends to scrape the mating counterface. However, C-E filled with $MoS_2$ indicated improved abrasion resistance under similar test conditions. This is attributed to a positive rolling effect of $MoS_2$ particles.

#### 9.3.2.2 Specific Wear Rate

The specific wear rate ($K_s$) of unoccupied C-E and $MoS_2$ and $Al_2O_3$ filled C-E composites at 23 and 34 N loads are presented in Figures 9.22a,b respectively. The $K_s$ data reveals that the $K_s$ tend to reduce with increasing abrading distance from 810 to 1080 m. However with increase in load from 23 to 34 N, they show increased $K_s$ and finally reaching a saturation level in all the cases. It was observed that wear behavior of C-E improved with the addition of silane treated $MoS_2$ and $Al_2O_3$ fillers. However, the higher loading of fillers in the C-E composite (10 wt% $MoS_2$ and $Al_2O_3$) leads to lower $K_s$ at all the loads and abrading distances. The maximum $K_s$ of unoccupied C-E and $Al_2O_3$ and $MoS_2$ filled C-E are $2.74 \times 10^{-11}$,

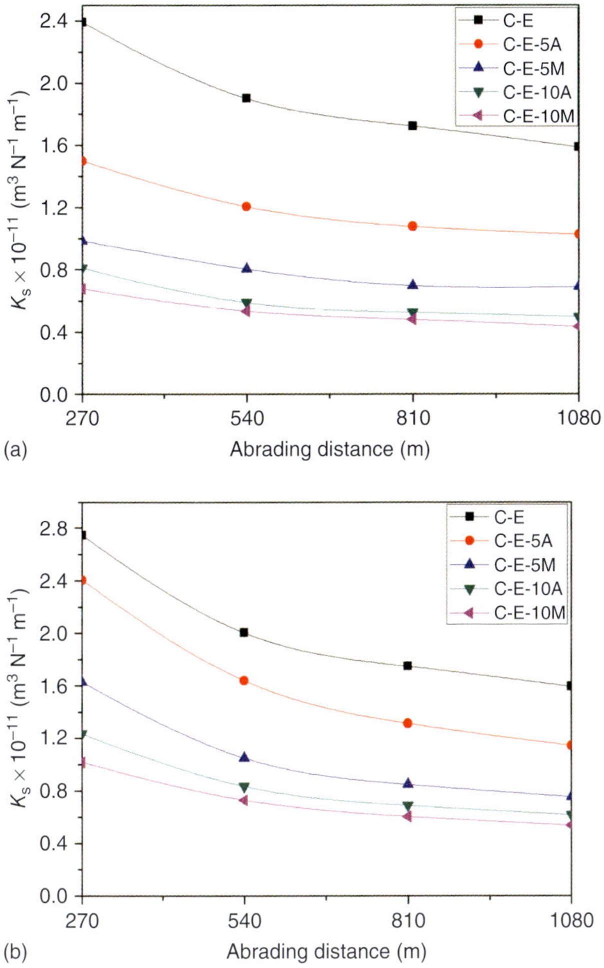


**Figure 9.22** $K_s$ versus abrading distance of particulate filled C-E composites at (a) 23 N and (b) 34 N.

$1.23 \times 10^{-11}$, and $1.01 \times 10^{-11}$ $m^3 N^{-1} m^{-1}$ respectively. The reason for the improvement in wear characteristics may be attributed to the type, surface treatment, and filler loading.

#### 9.3.2.3 Consequences of Factors on Wear Volume Loss

It can be observed that wear volume loss increases with increase in applied load and abrading distance while it decreases with increase in filler loading. At higher loads the composite material under investigation is fractured while with increase in abrading distance, it will be exposed to prolonged interaction with the abrasive particles, which contributes to the increase in wear volume [54]. The decrease in wear volume loss with increase in filler stacking is due to the silane coupling agent, which enriches the interfacial bonding with the polymer and fiber [55]. The type of particulate filler material has less significance than other factors on wear volume. Among the two filler materials used in the present study, $MoS_2$ leads to less wear volume loss compared to $Al_2O_3$. This may be attributed to the encouraging rolling outcome of $MoS_2$ particles during the abrasive wear process [41]. The $MoS_2$ particles seem to be pulled out by sand units along with matrix material during wear. These act as a rolling third body between the rubber wheel and the polymer composite, as shown in Figure 9.23. The $MoS_2$ elements behave like a ball-bearing element, meaning that the elements trundle relatively than glide among the two contacting surfaces, thereby reducing shear stress and $\mu$. This consequence of fillers has been previously mentioned in the works as "rolling effect" [56, 57], which is a pseudo-lubricant effect.

#### 9.3.2.4 Worn Surface Morphology

To associate the wear information effectively, SEM photomicrograph of the silica sand abrasives, scanned wear scar of selected samples, and worn surface heights of samples by vision measure scope (VMS, Mitutoyo make) are presented in Figures 9.24–9.26 respectively. The SEM photomicrographs of the damaged surfaces of unoccupied and particulate filled C-E samples are depicted in Figures 9.27–9.29 respectively. These specimens are exposed to an applied load of 34 N and abrading distance of 1080 m at a sliding velocity of 2.17 m s$^{-1}$. In the current discussion, silica sand (density 2.6 g cm$^{-3}$ and Knoop hardness 875) was cast off as the abrasive. The abrasive specks of American Foundry Society (AFS) grade silica sand were jagged in shape with pointed edges (Figure 9.24).

The distinctive wear damage of $MoS_2$ and $Al_2O_3$ stuffed C-E composite specimen is illustrated in Figure 9.25a,b respectively. Three different regions, i.e. entrance (denoted as 1), center (indicated as 2), and exit (denoted as 3), can be viewed under 3-BAW test at 34 N, 1080 m abrading distance. At the entrance and exit zones, the pressure on the abrasive is least, and the impairment morphologies are stable with particle rolling. The wear information and Figure 9.27 indicate comparatively soaring early wear rates when surfaces are fresh, and these slowly decrease with increase in abrading distance.

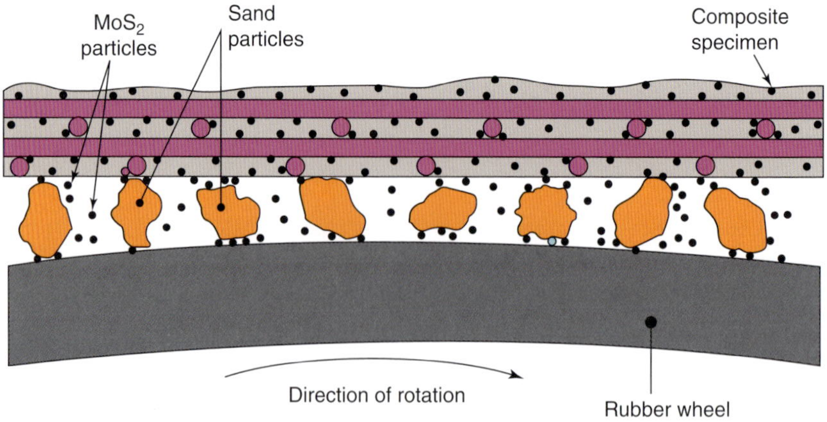


**Figure 9.23** Schematic representation of rolling effect of sand particles by $MoS_2$ microparticles.

**Figure 9.24** Scanning electron micrograph of silica sand before abrasion.

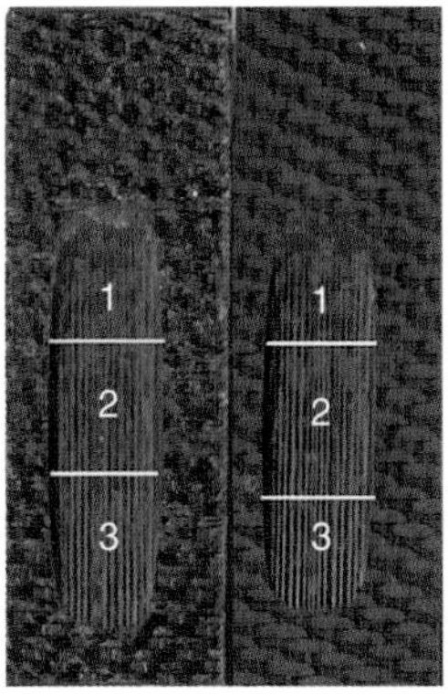


(a) (b)

**Figure 9.25** Wear scar of particulate filled samples at 34 N, 1080 m: (a) $Al_2O_3$ filled and (b) $MoS_2$ filled C-E composites.

This is owed to resin rich at the outer surface of the specimen, being rapidly worn until reinforcements resulting in the formation of stable surface. The wear volumes of unoccupied C-E and $Al_2O_3$ and $MoS_2$ filled C-E composite specimen at 34 N and 1080 m abrading distance are $0.567 \times 10^3$, $0.2189 \times 10^3$, and $0.1912 \times 10^3$ $mm^3$ respectively. To associate the wear volume damage with the depth of wear of dissimilar composites, the disparity in the depth of wear from entry to exit of the wear mark is illustrated in Figure 9.25. These illustrations indicate the ploughing marks in the abrading track and the length of the wear mark for unoccupied and particulate filled C-E specimens. The length of wear mark is more for the unoccupied C-E (41.4 mm) and least for $MoS_2$ filled C-E (35.95 mm) composite (Figure 9.25). Further, the wear scar of $MoS_2$ filled C-E is smoother (less coarse) than that of unoccupied and alumina filled C-E samples. The surface and nano-tribological behaviors were examined by XPS, atomic force microscopy (AFM), and nano-scratching experiments by Zhang et al. [58]. The surface

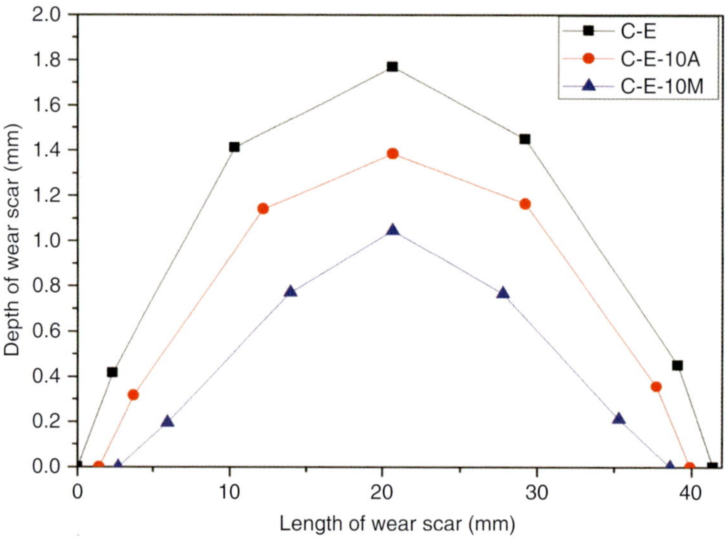


**Figure 9.26** Wear scar data of unoccupied C-E and particulate C-E samples.

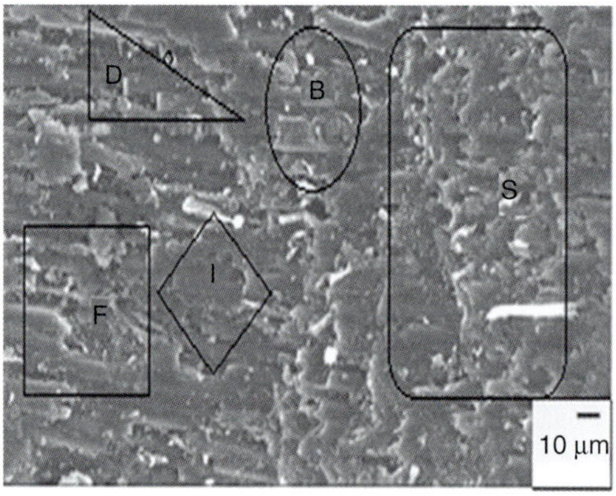


**Figure 9.27** Photomicrograph of damage surface of C-E sample at 34 N, 1080 m.

geography was investigated by surface imaging. High- and low-order roughness was decided by AFM categorization and was related to the diamond-like carbon (DLC) evolution mechanism and revealed the flattening effect of silver. They discovered that bonding friction was the leading friction mechanism and concluded that the bonding force between the scraping tip and DLC surface was diminished by hydrogenation and amplified by silver doping. In the present work, the surface geography was investigated by scanning and wear scar depth by VMS. The damaged surfaces exhibited better bonding between matrix and fiber/filler and smoothing effect of molybdenum disulfide than that of alumina.

Figure 9.27 demonstrates the damaged surface topographies of the unoccupied C-E composite specimen. The photomicrograph indicates that there is severe damage on the worn surface. The carbon fiber reinforced epoxy matrix

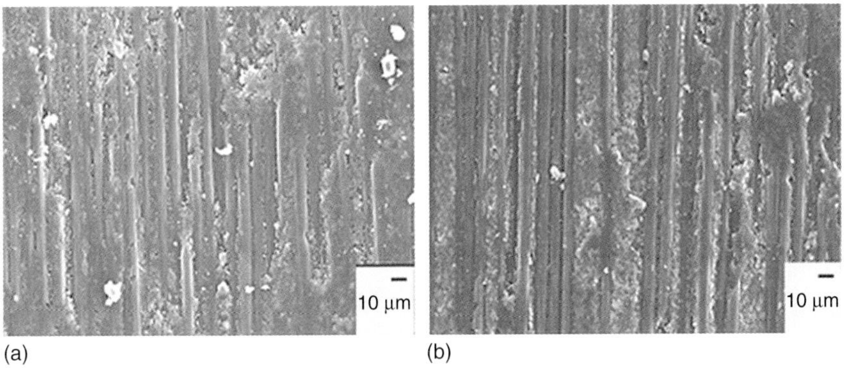


**Figure 9.28** Photomicrograph of worn surface of (a) $MoS_2$ filled C-E sample at 34 N, 270 m and (b) $MoS_2$ filled C-E sample at 34 N, 1080 m.

(Figure 9.27) was impaired more harshly than the particulate filled C-E composites (Figures 9.28 and 9.29). Figure 9.27 shows that the matrix is deformed and damaged (marked as D) by ploughing and the cutting action by jagged silica abrasive specks and uncovering of fibers. These uncovered fibers are inclined to rupture (marked as F) and get removed from the worn surface of the C-E composite. The micrograph also specifies crevice proliferation of the matrix, worsening of the fiber–matrix bonding owing to cyclic mechanical stress, and some fiber withdrawal from the matrix. Additional features such as the paced presence of carbon fibers (marked as S), fiber–matrix debonding (marked as B), and leaning end breakage of fibers (marked as I) can be seen from the photomicrograph in Figure 9.27. The breakage of fiber is due to abrasion and crosswise bending by pointed abrasive specks, resulting in breaking of fibers torn from the matrix (Figure 9.28).

Conversely, the SEM topographies of $MoS_2/Al_2O_3$ filled C-E samples are shown in Figures 9.28 and 9.29. It is apparent from the SEM photomicrographs (comparing Figure 9.27 with Figures 9.28a,b and 9.29a,b) that the 10 wt% $MoS_2/Al_2O_3$ filled C-E exhibits lower degree of damaged surface topographies

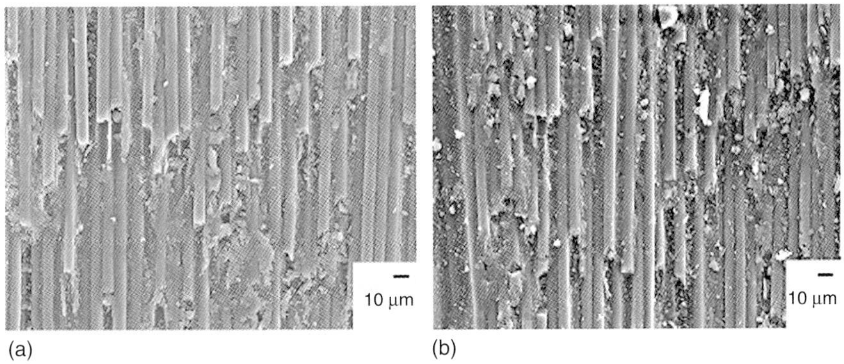


**Figure 9.29** Photomicrograph of worn surface of (a) $Al_2O_3$ filled C-E sample at 34 N, 270 m and (b) $Al_2O_3$ filled C-E sample at 34 N, 1080 m.

compared with unoccupied C-E specimen. These composites with abrasion examination possess lower rough surfaces, preserving the ploughing path along the abrading route (Figure 9.25). Wear damages emerge in the shape of lumped mass, indicating that they were created by micro-cutting. In addition to micro-cutting, micro-ploughing and micro-cracking appear in the worn surface. The abrasion is highly dependent on the asperity geometry. Zhang and Komvopoulos [59] studied the nanoscale pseudo elastic behavior of Cu-Al-Ni shape memory alloy (SMA) induced by partial indentation unloading and confirmed the direct implications in micro-system fabrication with SMAs. To have uniform surface roughness for all specimens, before abrasive wear test, the specimens were pre-worn using 1200 grit SiC paper and cleaned with a wire brush. Subsequent to the normal test duration, the sample was detached, meticulously gutted, and again weighed for accurate weight loss measurement. These topographies are important for the abrasive wear information, wherein the level of wear volume damage is elevated for the unoccupied C-E associated to the particulate filled C-E samples.

## 9.4 Conclusions

The role of ceramic and lubricating micro-fillers on friction and wear behavior of C-E composites has been evaluated by adhesive and abrasion wear processes. Unoccupied C-E composites were unfavorable to wear under upper normal loads and sliding velocities. Amalgamation of ceramic and lubricating fillers appreciably improved the wear performance of the C-E composite. Wear volume damage and friction coefficient of all the composites increased with increase in the normal load/sliding velocity. $Al_2O_3$ filled C-E composites exhibited slightly higher wear volume loss compared to $MoS_2$ filled C-E composites. Lower friction coefficient was observed in 10 wt% $MoS_2$ filled C-E composites, which demonstrated better resistance to wear. $Al_2O_3$ filled C-E composites exhibited slightly higher friction coefficient compared to $MoS_2$ filled and unoccupied C-E composites. The wear mechanism of the unfilled C-E composite during dry sliding wear was dominated by plastic deformation, fiber thinning, bending of transverse fiber, and fiber breakage, whereas in filled composites, micro-cracking, fine debris formation, less fiber breakage, and a thin continuous transfer film on the counterface were the principal wear mechanisms. Lastly, it can be concluded that C-E composites exhibited superior wear presentation when filled with a lubricating filler, namely $MoS_2$.

The experimental outcomes of 3-BAW showed increase in $K_s$ with increase in normal load while it decreased with increase in abrading distance for unfilled as well as particulate filled C-E composites. Also, particulate filled C-E composites performed better in contrast with unoccupied C-E composite. The micrographs of three-body abrasive worn surface features exhibit matrix and fiber damages, fiber breakage, and fiber elimination, the magnitude of which changed according to the system, load, type of filler material, and abrading distance. The chief wear mechanisms in particulate filled C-E composites are micro-cutting, micro-ploughing, and micro-cracking.

## References


1 Ratna, D., Manoj, N.R., and Varley, N.R. (2003). Clay-reinforced epoxy nanocomposites. *Polymer International* 52 (9): 1403–1407.
2 DeAlbuquerque, A., Joseph, K., Hecker de Carvalho, L., and Morais D'Almeida, J. (1999). Effect of wettability and ageing conditions on the physical and mechanical properties of uniaxially oriented jute-roving-reinforced polyester composites. *Composite Science and Technology* 60: 833–844.
3 Devi, L., Bhagawan, S., and Thomas, S. (1997). Mechanical properties of pineapple leaf fiber-reinforced polyester composites. *Journal of Applied Polymer Science* 64: 1739–1748.
4 Mwaikambo, L. and Bisanda, E. (1999). The performance of cotton/kapok fabric-polyester composites. *Polymer Testing* 18: 181–198.
5 Maldas, D. and Kokta, B.V. (1995). Composites of chlorinated polyethylene wood fibers. *Journal of Reinforced Plastic Composites* 14: 458–470.
6 Sain, M.M. and Kokta, B.V. (1994). Polyolefin-wood filler composite. I. Performance of *m*-phenylene bismaleimide-modified wood fiber in polypropylene composites. *Journal of Applied Polymer Science* 54: 1545–1559.
7 Mallik, P.K. (2007). *Fiber-Reinforced Composite Materials, Manufacturing and Design*. London: CRC Press.
8 Saegusa, T. (1995). Organic–inorganic polymers hybrids. *Pure Applied Chemistry* 67: 1965–1970.
9 Gao, Y. and Choudhury, N.R. (2003). Organic–inorganic hybrid from ionomer. In: *Handbook of Organic–Inorganic Hybrid Materials and Nanocomposites* (ed. H.S. Nalwa), 271–295. American Scientific Publisher.
10 Schwartz, M.M. (1992). *Composite Materials Handbook*. New York: McGraw-Hill.
11 Raj, R.G., Kokta, B.V., and Daneault, C.A. (1990). A comparative study on the effect of ageing on mechanical properties of LDPE-glass fiber, mica, and wood fiber composites. *Journal of Applied Polymer Science* 40: 645–655.
12 Ravi Kumar, B.N., Suresha, B., and Venkataramareddy, M. (2009). Effect of particulate fillers on mechanical and abrasive wear behaviour of polyamide 66/polypropylene nanocomposites. *Materials and Design* 30 (9): 3852–3858.
13 Hemanth, R., Suresha, B., and Sekar, M. (2015). Physico-mechanical behaviour of thermoplastic co-polyester elastomer/polytetrafluoroethylene composite with short fibers and microfillers. *Journal of Composite Materials* 49 (18): 2217–2229.
14 Harsha, A.P. and Tewari, U.S. (2003). The effect of fibre reinforcement and solid lubricants on abrasive wear behavior of polyetheretherketone composites. *Journal of Reinforced Plastics Composites* 22: 751–767.
15 Subbaya, K.M., Suresha, B., Rajendra, N., and Varadarajan, Y.S. (2012). Grey-based Taguchi approach for wear assessment of SiC filled carbon-epoxy composites. *Materials and Design* 41: 124–130.
16 Ramesh, B.N. and Suresha, B. (2014). Optimization of tribological parameters in abrasive wear mode of carbon-epoxy hybrid composites. *Materials and Design* 59: 38–49.

17 Chauhan, S.R., Kumar, A., Singh, I., and Kumar, P. (2010). Effect of fly ash content on friction and dry sliding wear behavior of glass fiber reinforced polymer composites – a Taguchi approach. *Journal of Minerals and Materials Characterization and Engineering* 9 (4): 365–387.

18 Siddhartha, A.P. and Bhatt, A.D. (2011). Mechanical and dry sliding wear characterization of epoxy-$TiO_2$ particulate filled functionally graded composites materials using Taguchi design of experiment. *Materials and Design* 32 (2): 615–627.

19 Akram, W., Chaturvedi, S.K., and Ali, S.M. (2013). Comparative study of mechanical properties of e-glass/epoxy composite materials with $Al_2O_3$, $CaCo_3$, $SiO_2$ and PBO fillers. *International Journal of Engineering Research and Technology* 2 (7): 1029–1034.

20 Kuljanin, J., Vuckovic, M., Comor, M.I. et al. (2002). Influence of CdS-filler on the thermal properties of polystyrene. *European Polymer Journal* 38 (8): 1659–1662.

21 Weidenfeller, B., Hofer, M., and Schilling, F. (2004). Thermal conductivity, thermal diffusivity and specific heat of particle filled polypropylene. *Composites Part A: Applied Science and Manufacturing* 35 (4): 423–429.

22 Bijwe, J. and Indumathi, J. (2004). Influence of fibers and solid lubricants on low amplitude oscillating wear of polyetherimide composites. *Wear* 257: 562–572.

23 Neale, M.J. and Gee, M. (2001). *Guide to Wear Problems and Testing for Industry*. New York: William Andrew Publishing.

24 Zum Gahr, K.H. (1998). Wear by hard particles. *Tribology International* 31: 587–596.

25 Gates, J.D. (1998). Two-body and three-body abrasion: a critical discussion. *Wear* 214: 139–146.

26 Chand, N., Naik, A.M., and Neogi, S. (2000). Three-body abrasive wear of short glass fiber polyester composite. *Wear* 242 (1–2): 38–46.

27 Suresha, B. and Chandramohan, G. (2008). Three-body abrasive wear behavior of particulate-filled glass-vinyl ester composites. *Journal of Material Processing and Technology* 200 (1): 306–311.

28 Suresha, B., Chandramohan, G., Kishore et al. (2008). Mechanical and three-body abrasive wear behavior of SiC filled glass-epoxy composites. *Polymer Composites* 29 (9): 1020–1025.

29 Harsha, A.P. and Tewari, U.S. (2003). Two-body and three-body abrasive wear behavior of polyaryletherketone composites. *Polymer Testing* 22 (4): 403–418.

30 Suresha, B., Chandramohan, G., and Mohanram, P.V. (2009). Role of fillers on three-body abrasive wear behavior of glass fabric reinforced epoxy composites. *Polymer Composites* 30 (8): 1106–1113.

31 ASTM G99-17 (2017). Standard test method for wear testing with a pin-on-disk apparatus. West Conshohocken, PA: ASTM International.

32 ASTM G65-16e1 (2016). Standard test method for measuring abrasion using the dry sand/rubber wheel apparatus. West Conshohocken, PA: ASTM International.

33 Xing, X.S. and Li, R.K.Y. (2004). Wear behaviour of epoxy matrix composites filled with uniform sized sub-micron spherical silica particles. *Wear* 256: 21–26.

34 Wang, Y. and Yan, F. (2006). Tribological properties of transfer films of PTFE-based composites. *Wear* 261: 1359–1366.
35 Lansdown, A.R. (1992). *Lubrication – A Practical Guide to Lubricant Selection*. London: Pergamon Press.
36 Brudnyi, A.I. and Karmadonov, A.F. (1975). Structure of molybdenum disulphide lubricant film. *Wear* 33 (2): 243–249.
37 Shankara, A., Menezes, P.L., Simha, K.R.Y., and Kailas, S.V. (2008). Study of solid lubrication with $MoS_2$ coating in the presence of additives using reciprocating ball-on-flat scratch tester. *Sadhana* 33 (3): 207–220.
38 Li, X., Gao, Y., Xing, J. et al. (2004). Wear reduction mechanism of graphite and $MoS_2$ in epoxy composites. *Wear* 257: 279–283.
39 Kishore and Kumar, K. (1998). Sliding wear studies in epoxy containing alumina powders. *High Temperature Materials Processing* 17: 271–274.
40 Durand, J.M., Vardavoullas, M., and Jeandin, M. (1995). Role of reinforcing ceramic particles in the wear behaviour of polymer-based model composites. *Wear* 181–183: 833–839.
41 Pettarin, V., Churruca, M.J., Felhos, D. et al. (2010). Changes in tribological performance of high molecular weight high density polyethylene induced by the addition of molybdenum disulphide particles. *Wear* 269: 31–45.
42 Wang, J., Gua, M., Songhao, B., and Gec, S. (2003). Investigation of the influence of $MoS_2$ filler on the tribological properties of carbon fiber reinforced Nylon 1010 composites. *Wear* 255: 774–779.
43 Suresha, B., Chandramohan, G., Prakash, J.N. et al. (2006). The role of fillers on friction and slide wear characteristics in glass-epoxy composite systems. *Journal of Minerals and Materials Characteristics Engineering* 5: 87–101.
44 Suresha, B., Siddaramaiah, Kishore et al. (2009). Investigations on the influence of graphite filler on dry sliding wear and abrasive wear behaviour of carbon fabric reinforced epoxy composites. *Wear* 267: 1405–1414.
45 Tsukizoe, T. and Ohmae, N. (1983). Friction and wear of advanced composite materials. *Fiber Science and Technology* 18: 265–286.
46 Bijwe, J. and Satapathy, B.K. (2002). Analysis of simultaneous influence of operating variables on abrasive wear of phenolic composites. *Wear* 253: 787–794.
47 Lancaster, J.K. (1972). Friction and wear. In: *Polymer Science: A Material Science Hand Book* (ed. A.D. Jenkins), 959–1046. Amsterdam: North-Holland.
48 Schwartz, C.J. and Bahadur, S. (2000). Studies on the tribological behaviour and transfer film-counterface bond strength for polyphenylene sulphide filled with nanoscale alumina particles. *Wear* 237: 261–273.
49 Patnaik, A., Satapathy, A., and Biswas, S. (2010). Investigation on three-body abrasive wear and mechanical properties of particulate filled glass epoxy composites. *Malaysian Polymer Journal* 5: 37–48.
50 Bahadur, S. and Gong, D.L. (1992). The action of fillers in the modification of the tribological behaviour of polymers. *Wear* 158: 41–59.
51 Winer, W.O. (1967). Molybdenum disulphide as a lubricant: a review of fundamental knowledge. *Wear* 10: 422–452.
52 http://www.ferret.com.au/c/Dotmar-EPP/Ertalon-and-Nylatron-from-Dotmar-EPP-n735215 (accessed 15 October 2017).

53 Sole, B.M. and Ball, A. (1996). On the abrasive wear behaviour of mineral filled polypropylene. *Tribology International* 29: 457–465.

54 Rajeev, V.R., Dwivedi, D.K., and Jain, S.C. (2011). A fractional factorial design study of reciprocating wear behaviour of Al-Si-$SiC_p$ composites at lubricated contacts. *Journal of Materials Engineering Performances* 20: 368–376.

55 Suresha, B., Ramesh, B.N., Subbaya, K.M., and Chandramohan, G. (2010). Mechanical and three-body abrasive wear behaviour of carbon-epoxy composite with and without graphite filler. *Journal of Composite Materials* 44: 2509–2519.

56 Jiang, Z., Gyurova, L.A., Schlarb, A.K. et al. (2008). Study on friction and wear behavior of polyphenylene sulfide composites reinforced by carbon fibers and sub-micro $TiO_2$ particles. *Composite Science and Technology* 68: 734–742.

57 Rapoport, L., Lvovsky, M., Lapsker, I. et al. (2001). Slow release of fullerene-like $WS_2$ nanoparticles as a superior solid lubrication mechanism in composite matrices. *Advanced Engineering Materials* 3: 71–75.

58 Zhang, H.S., Endrino, J.L., and Anders, A. (2008). Comparative surface and nano-tribological characteristics of nanocomposite diamond-like carbon thin films doped by silver. *Applied Surface Science* 295: 2551–2556.

59 Zhang, H.S. and Komvopoulos, K. (2010). Nanoscale pseudoelastic behaviour of Cu-Al-Ni shape-memory alloy induced by partial indentation unloading. *Nanoscience Nanotechnology Letters* 2: 332–336.

# 10

# Dry Sliding Wear Behavior of Copper Based Hybrid Metal Matrix Composite

*Ponnambalam Balamurugan, Marimuthu Uthayakumar, and Sundaresan Thirumalai Kumaran*

*Kalasalingam University, Department of Mechanical Engineering, Anand Nagar, Krishnankoil 626126, Tamil Nadu, India*

## 10.1 Introduction

In the present day scenario, the development of metal matrix composites (MMCs) has been receiving global recognition because of its superior properties in terms of strength, stiffness, wear resistance, and corrosion resistance compared to monolithic metals. MMCs are generally formed by the blend of a hard reinforcement phase in the metallic matrix. The different matrix materials used are copper, magnesium, aluminum, and their alloys [1–4]. Alumina, silicon carbide, titanium carbide, tungsten carbide, carbon nanotubes, boron nitrate, graphite, and molybdenum disulfide are the reinforcement materials used widely along with the metal matrix to form MMCs [5–11]. In the ductile matrix the stress concentrations and sudden propagation of cracks are reduced because of plastic deformation behavior, which enables improvement in fracture toughness. Applications such as aerospace, automobile, marine, and turbine-compressor engineering use components made of MMCs in the recent years, because of their low density, greater strength and stiffness, and resistance to wear and high temperature [12–17]. In the automobile industry MMCs are used to manufacture components such as brake disk and drums, cylinder blocks, and liners drive shaft [18, 19]. The aircraft industry uses MMCs to manufacture components such as rotor vanes, rotor blades, and drive shafts.

Copper and its alloys are widely used in various industries due to the significant properties such as good thermal and electrical conductivity, corrosive resistance, nonmagnetic, and stability at elevated temperatures [20, 21]. Copper has higher thermal conductivity, which is an important property for making components requiring fast heat dissipation; however, properties such as high deformability and low strength make copper inappropriate for many industrial applications [22]. To increase the strength, wear, and frictional resistance, an additional reinforcement needs to be added to the copper matrix. Moustafa et al. [23] reported in their study that high electrical conductivity and excellent

*Synthesis and Tribological Applications of Hybrid Materials*, First Edition.
Edited by Mohammad Jawaid, Rajini Nagarajan, Jacob Sukumaran, and Patrick De Baets.

lubrication properties render the copper–graphite composites useful in the manufacture of sliding contacts. Kalinin et al. [24] reported that copper based composites have applications in making components such as connectors, electrodes for spot welding processes, lead wires, and other electronic components because of the exclusive combination of strength and conductivity at high temperatures. Kato et al. [25] elaborated that copper based composites made by powder metallurgy are commonly used for making components such as bushes and bearings. Copper–tin composites have been developed to make self-lubricating bearings working in extreme load and temperature conditions. Hussain and Kit [26] investigated the copper–alumina composites prepared through mechanical alloying and ball milling for studying the spot welding behavior and mechanical properties; their results revealed that a mechanically alloyed composite shows good performance on spot welding behavior, since the electrical current is constrained within a small weld region due to higher hardness achieved through mechanical alloying. Hong et al. [27] manufactured a composite with copper alloy as matrix and tungsten carbide as reinforcement with or without indium for the application of drill bit in oil or gas industry. Ram Prabhu et al. [28] reported the potential use of copper based composites for applications such as clutches and brakes and recommended copper hybrid composites prepared with SiC and graphite as reinforcements for brake friction applications.

Fly ash is a waste by-product obtained from thermal power plants that use coal as the primary fuel for power production. Fly ash is available at low cost and has low density, which makes it suitable for use as reinforcement in MMCs. Fly ash is one of the cheapest and low-density reinforcements obtained in large quantities as a waste by-product during combustion of coal in thermal power plants, which can be utilized to develop metal matrix composites [29] . Some researchers recommended fly ash cenosphere particles for foam based applications [30, 31]. Owing to its low cost and low density, fly ash can be used as reinforcement in metal matrix [32, 33]. Since it is an underutilized by-product from burning of coal from the thermal power [34], reduction in environmental pollution created due to land filling can be achieved by utilizing it. Mahendra and Radhakrishna [35] studied the characterization of stir cast Al–Cu–(fly ash + SiC) hybrid metal matrix composites. They conveyed that the addition of fly ash decreases the density and increases the tensile strength, compression strength, impact strength, and wear resistance. Kumar et al. [36] performed the dry sliding wear behavior study at elevated temperatures on AA6061 to find out the effect of addition of particulate fly ash on the aluminum alloy. The composite was manufactured through powder metallurgy route. They proved that addition of fly ash on AA6061 delivered better wear behavior at both room and elevated temperatures. They stated that incorporation of fly ash improved the wear resistance up to 300 °C due to hardening effect on the surface by the fly ash particles. David Raja Selvam et al. [37] developed the AA6061 aluminum alloy composites by the compo-casting technique; they reported the microstructure and some mechanical properties of fly ash particulate reinforcement in the aforesaid alloy. Their results revealed that the addition of fly ash particles enhanced the micro hardness and tensile strength of the AMCs. AA6061/12 wt%

fly ash composite showed 132.21% higher micro hardness and 56.95% higher ultimate tensile strength compared to unreinforced AA6061 alloy. Sreekanth et al. [38] explained the effect of addition of fly ash on polymer based composites with varying particle size and reinforcement percentage. They reported that addition of fly ash improved the strength and thermal stability; on the contrary, elongation at break decreases drastically. They claimed significant improvement in mechanical properties that was found in the composites prepared with smaller particle size of fly ash. Sai et al. [39] reported that with the addition of fly ash up to 8 wt%, the wear resistance of the composite increases. Vogiatzis et al. [40] prepared syntactic foams comprising of aluminum and ceramic cenospheres with different densities through powder metallurgy technique by varying the compaction pressure, ranging from 200 to 300 MPa. From the compression study, they claimed that fracture propagation is from the bottom to top, detachment of the external wall is from the main body, and the top surface of the specimen is not damaged; shear planes are formed in the core of the specimens. Fan et al. [41] explained the reaction effect between the fly ash and Al–Mg melt for different holding times during the stir casting process. The considered holding time varies from 10 to 40 h. Their results show that for longer reaction time the fly ash decomposed and generated θ-alumina and $MgAl_2O_4$. $SiO_2$ is converted to Si, and the magnesium alloy reacts with the Si to form $Mg_2Si$. When the reaction time is more than 30 h nearly all the fly ash in the material decomposes.

According to Rao et al. [42] stir casting method is preferred because of its ease, flexibility, and suitability to large scale production. They also stated that the only problem associated with the stir casting process is the distribution of the particulate in a non-uniform manner, due to segregation of particles by gravity and low wettability. Sajjadi et al. [43] reported that because of the better wettability of particles in compo-casting technique than in the stir casting, higher densification and lower grain size are achieved in compo-casting. Sapate et al. [44] reported that for making copper matrix composite, powder metallurgy technique is appropriate, since copper has a high melting point, and also possesses poor wettability with the reinforcement particles. Narayanasamy et al. [45] reported that powder metallurgy route is preferred for making materials such as cutting tools made of tungsten carbide, and nickel based super alloys blended through mechanical alloying. Powder metallurgy is popular globally due to its high productivity, ability to make intricate shapes with precision, and cost-effectiveness on mass production. Ayyappadas et al. [46] fabricated the graphene reinforced copper matrix composites with varying proportions through powder metallurgy technique. They suggested graphene as the solid lubricant for MMCs in dry sliding conditions, since the friction coefficient becomes low with increase in graphene content. They also reported that microwave sintered composites show better performance than the conventional sintered composites due to higher densification.

Rajkumar and Aravindan [47] prepared a hybrid copper based composite with titanium carbide and graphite as reinforcements through microwave sintering and studied its performance on tribological characteristics. They stated that wear rate and coefficient of friction (CoF) of hybrid composites and unreinforced copper increase with increase in normal load. Unreinforced copper has higher wear rates and CoF compared to hybrid composites. Formation of oxide layer in the contact surface and deformation of the microstructure occur, which signifies higher wear rate on unreinforced copper. Jia et al. [48] studied the wear behavior and friction characteristics of the bronze graphite composites. The studies were carried out in dry sliding condition and water lubricated condition. Stainless steel was used as the counter surface material for the wear studies. They reported that during dry sliding condition plastic deformation, micro-cracking, and scuffing were attributed to adhesion and abrasive wear behavior of the composite. They also reported the transfer of Fe from the counter face to the specimen under dry sliding condition, which is not noticed in the specimen tested under lubricated condition.

In the present chapter, dry sliding wear behavior of copper fly ash graphite hybrid composites with varying graphite percentage is studied under different sliding velocities and loads.

## 10.2 Materials and Methods

### 10.2.1 Materials

Copper metal powder with 99.7% purity with an average particle size of 30 μm was purchased from the Oxford Chemicals, which was used as the matrix. Copper is a chemical element with symbol Cu (from Latin: cuprum) and atomic number 29. It is a soft, malleable, and ductile metal with very high thermal and electrical conductivity. The melting point of copper is 1085 °C and the density is 8940 kg $m^{-3}$. A freshly exposed surface of pure copper has a reddish-orange color. The purchased copper powder is shown in Figure 10.1.

Fly ash was used as the reinforcement material. Fly ash is one of the residues generated in combustion and the fine particles rise with the flue gases as flue particles. Fly ash is usually referred to the ash produced during combustion of

**Figure 10.1** Copper powder.

(a) (b)

**Figure 10.2** Morphology of (a) copper powder and (b) fly ash particles.

coal. The melting point of fly ash is 1400 °C and its density is 890 kg m$^{-3}$. The properties of fly ash are low density, high wear, and abrasion resistance. Fly ash obtained from Tuticorin Thermal Power Station, India, with an average particle size of 5 μm was used as reinforcement. The composition of the fly ash obtained is as follows: $SiO_2$ – 61.75%; $Al_2O_3$ – 25.24%; $Fe_2O_3 + Fe_3O_4$ – 4.77%; CaO – 1.3%; MgO – 0.87%. The morphology of copper and fly ash was obtained through scanning electron microscope (SEM), which is shown in the Figure 10.2. From the SEM images, it is found that copper powder is in the form of dendrite structure and fly ash is in the form of hollow spheres.

### 10.2.2 Preparation of the Composite by Powder Metallurgy Process

Three proportions of composites with copper as matrix and fly ash with 10 wt% along with 1, 2, and 3 wt% graphite were prepared by maintaining copper +10 wt% fly ash as the common proportion for all composites. The copper, fly ash, and graphite powders of required weight proportions were taken by weighing the powders using a digital weighing balance having 0.0001 g accuracy. The proportions taken were blended manually and heated to 100 °C under argon atmosphere to remove the moisture content before compacting. After heating, compaction was done using single action compaction die in the universal testing machine (UTM) at the compaction pressure of 450 MPa. Before compaction process, the inner surfaces of the die were cleaned with acetone, and after drying the surfaces were lubricated with wax to avoid the adhesion of the powder with the die and to enhance the removal of specimen from the die easily. The die along with the UTM during the compaction process is shown in the Figure 10.3.

The product obtained by the compaction process is called green compact. The prepared green compact specimen was of 20 mm diameter. In green compact the particles were only mechanically bonded. To make the particles metallurgically bonded, the prepared green compact was sintered using the tubular furnace under argon atmosphere at 900 °C for a period of 60 min. The sintered specimen was machined to the required dimension for further testing. The specimen after compacting is shown in the Figure 10.4.

**Figure 10.3** Compaction process.

**Figure 10.4** Typical specimen after compacting.

### 10.2.3 Wear Studies

The sintered specimen was machined to the dimension of 30 mm height and 10 mm diameter. Wear studies were carried out using pin-on-disk setup as per ASTM G99-2010. The prepared composite specimens were tested for the loads of 10, 30, and 50 N under the sliding velocities of 1, 2, and 3 $m\,s^{-1}$. Before conducting the experiment, the specimen was polished with 800, 1000, and 1200 grit

**Figure 10.5** Pin-on-disk setup.

**Figure 10.6** Typical wear specimen.

emery paper followed by polishing with alumina particles on a double disk polishing machine. The counter disk for wear testing was made of EN31 Steel having 65HRC and dimensions of 165 mm diameter and 8 mm thickness with the arithmetic average roughness (Ra) of 0.84 μm. Analysis of wear mechanism was carried out using SEM/EDX. The pin-on-disk setup is shown in Figure 10.5. The typical wear specimen is shown in Figure 10.6.

## 10.3 Results and Discussion

Figure 10.7 shows the microstructure of the composite with 1 wt% graphite. Figures 10.8 and 10.9 show the microstructures of the composite with 2 and 3 wt% graphite respectively. From the microstructure results it is inferred that reinforcements have a fairly uniform distribution in the matrix. The white background portions represent the copper matrix. The black rounded dot-like structure represents the fly ash and the black patches in non-uniform shape represent the graphite.

Figure 10.10a–c shows the wear test results of the different composites prepared under sliding velocities of 1, 2, and 3 $m\,s^{-1}$ in the loading condition of 10,

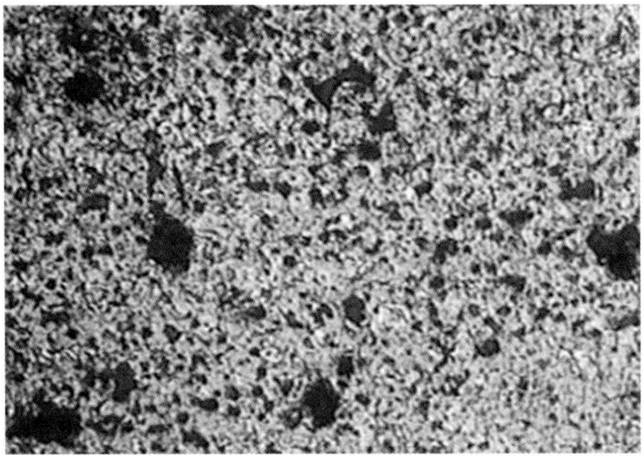

**Figure 10.7** Microstructure of composite with 10 wt% fly ash and 1 wt% graphite.

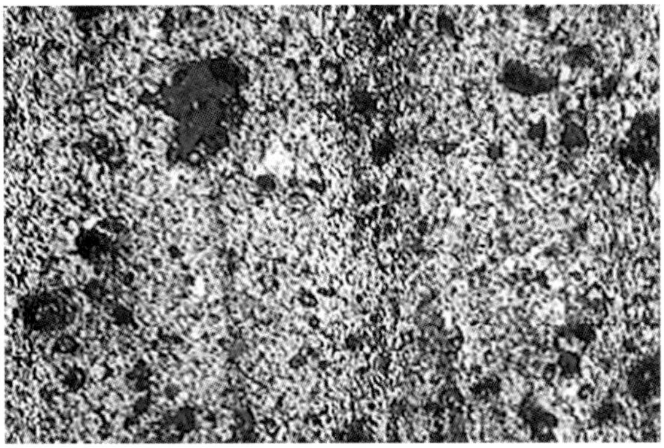

**Figure 10.8** Microstructure of composite with 10 wt% fly ash and 2 wt% graphite.

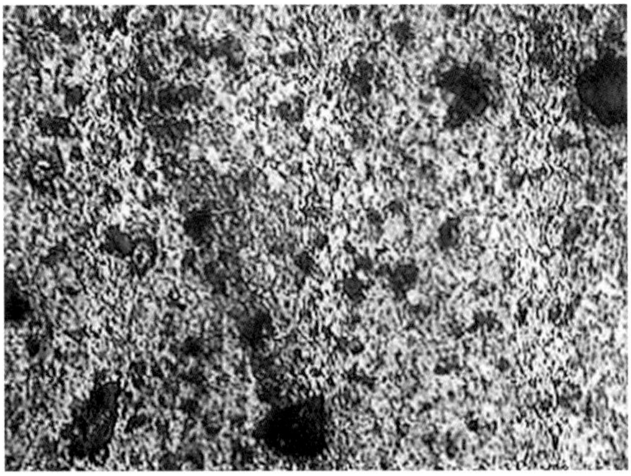

**Figure 10.9** Microstructure of composite with 10 wt% fly ash and 3 wt% graphite.

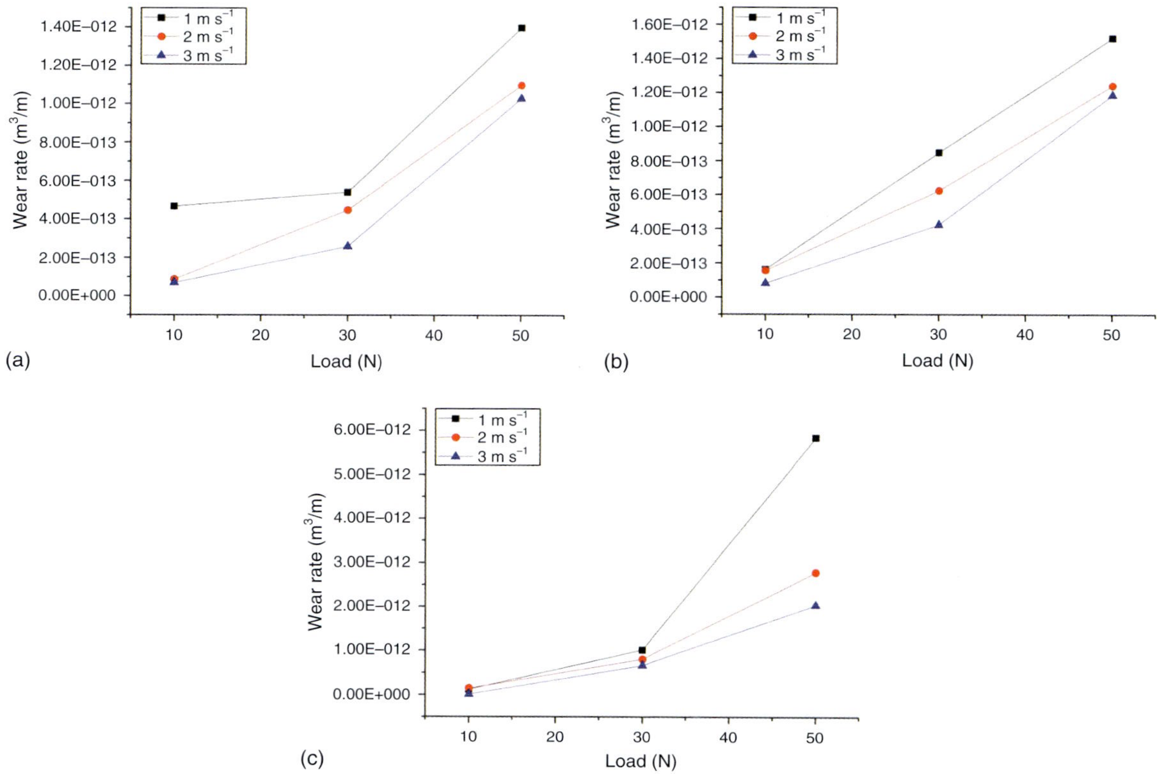


**Figure 10.10** Effect of load and sliding velocity on wear rate of the composite with (a) 10% fly ash and 1% graphite, (b) 10% fly ash and 2% graphite, and (c) 10% fly ash and 3% graphite.

30, and 50 N. Figure 10.10a shows the effect of load and sliding velocity on wear rate of the composite with 10% fly ash and 1% graphite. The results indicate that as the load increases the wear rate increases under all sliding velocities, or vice versa – for the same loading condition as the velocity increases the wear rate decreases; this is due to the decrease in CoF. As the sliding velocity increases CoF decreases due to the formation of a mechanically mixed layer between the sliding and the courter parts. Similar results are observed for the composites with higher graphite contents, which are shown in Figure 10.10b,c.

It is evident from Figure 10.11a–c that as the sliding velocity increases the CoF decreases for all composites prepared. The results also show that as the reinforcement percentage of graphite increases from 1% to 3% there is a reasonable decrease in CoF, which is achieved through increased lubrication layer formation with the increase in graphite content.

To find the influence of each input parameter (load, sliding velocity, and percentage reinforcement) on the output responses (wear rate and CoF) gray relational analysis was done. The gray relational analysis was done based on L27 orthogonal array. The input and output responses, signal-to-noise ratio, normalization gray relational coefficient, and gray relational grade are shown in Table 10.1. The signal-to-noise ratio for the output response is calculated based on Equation (10.1):

$$10\log\left(X^2\right) \tag{10.1}$$

where $X$ is the output response.

The normalized value is calculated using Equation (10.2):

$$Z_{ij} = \frac{\max\left(Y_{ij}\right) - Y_{ij}}{\max\left(Y_{ij}\right) - \min\left(Y_{ij}\right)} \tag{10.2}$$

where $Y_{ij}$ is the output response for the combination based on L27 orhtogonal array.

The preliminary values are calculated by finding the absolute difference between $Y_o(k)$ and $Y_{ij}$ where $Y_o(k) = 1$.

The gray relational coefficient is calculated using Equation (10.3):

$$\gamma\left(y_o(k), y_i(k)\right) = \frac{\Delta\min + \xi\Delta\max}{\Delta_{oj}(k) + \xi\Delta\max} \tag{10.3}$$

$\xi$ is assumed as 0.5.

Gray relational grade is calculated based on Equation (10.4):

$$\bar{\gamma}_j = \frac{1}{k}\sum_{i=1}^{m}\gamma_{ij} \tag{10.4}$$

where $\bar{\gamma}_j$ is the gray relational grade for the $j$th experiment, and $k$ is the number of output responses.

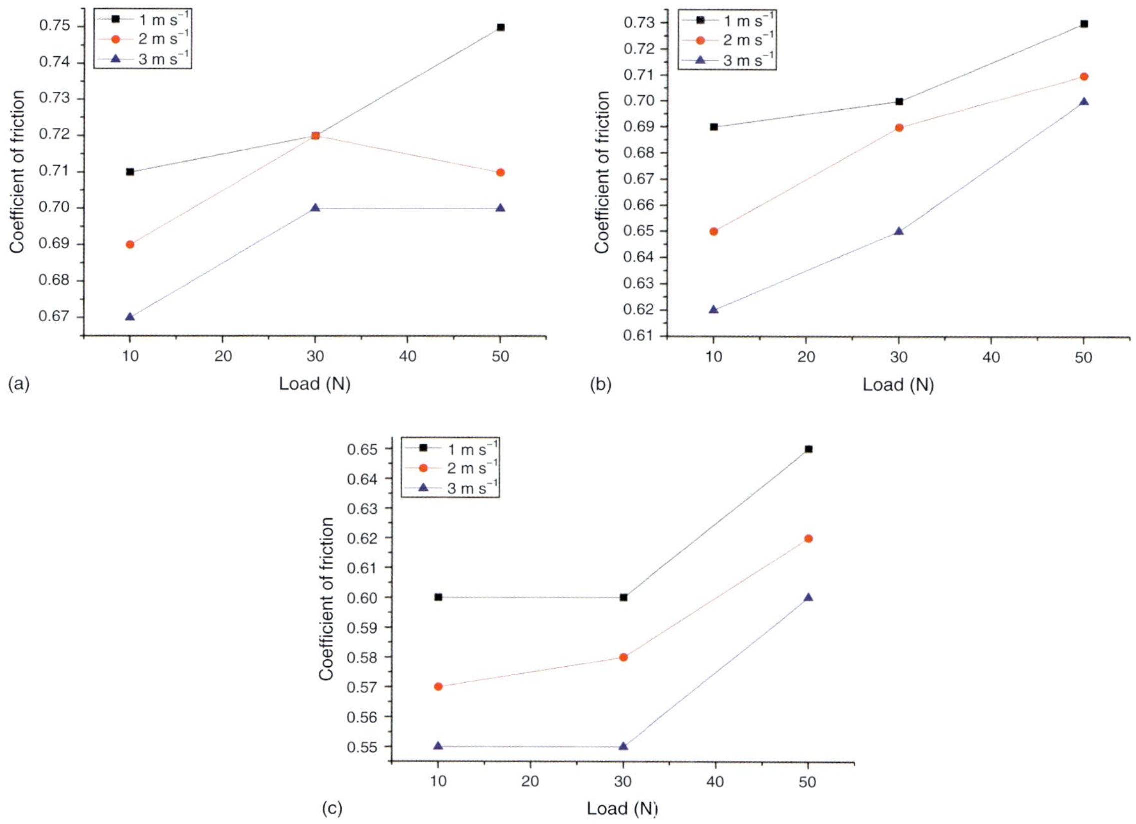


**Figure 10.11** Effect of load and sliding velocity on CoF of the composite with (a) 10% fly ash and 1% graphite, (b) 10% fly ash and 2% graphite, and (c) 10% fly ash and 3% graphite.

**Table 10.1** Grey relational analysis.

| S. No | Load | Sliding velocity | % graphite | Error | Wear rate | CoF | Signal-to-noise ratio (wear rate) | Signal-to-noise ratio (CoF) | Normalization (wear rate) | Normalization (CoF) | GRC (wear rate) | GRC (CoF) | GRG | Order |
|---|---|---|---|---|---|---|---|---|---|---|---|---|---|---|
| 1 | 10 | 1 | 1 | 1 | 4.66E−13 | 0.71 | 246.6323 | 2.9748 | 0.9220 | 0.2000 | 0.8651 | 0.3846 | 0.6248 | 15 |
| 2 | 10 | 1 | 2 | 2 | 1.62E−13 | 0.69 | 255.7842 | 3.2230 | 0.9741 | 0.3000 | 0.9507 | 0.4167 | 0.6837 | 12 |
| 3 | 10 | 1 | 3 | 3 | 1.22E−13 | 0.6 | 258.2389 | 4.4370 | 0.9809 | 0.7500 | 0.9632 | 0.6667 | 0.8150 | 4 |
| 4 | 10 | 2 | 1 | 2 | 8.67E−14 | 0.69 | 261.2329 | 3.2230 | 0.9870 | 0.3000 | 0.9747 | 0.4167 | 0.6957 | 10 |
| 5 | 10 | 2 | 2 | 3 | 1.56E−13 | 0.65 | 256.1249 | 3.7417 | 0.9751 | 0.5000 | 0.9526 | 0.5000 | 0.7263 | 7 |
| 6 | 10 | 2 | 3 | 1 | 1.46E−13 | 0.57 | 256.6994 | 4.8825 | 0.9768 | 0.9000 | 0.9557 | 0.8333 | 0.8945 | 3 |
| 7 | 10 | 3 | 1 | 3 | 6.86E−14 | 0.67 | 263.2744 | 3.4785 | 0.9902 | 0.4000 | 0.9807 | 0.4545 | 0.7176 | 8 |
| 8 | 10 | 3 | 2 | 1 | 8.12E−14 | 0.62 | 261.8048 | 4.1522 | 0.9880 | 0.6500 | 0.9766 | 0.5882 | 0.7824 | 5 |
| 9 | 10 | 3 | 3 | 2 | 1.12E−14 | 0.55 | 278.9860 | 5.1927 | 1.0000 | 1.0000 | 1.0000 | 1.0000 | 1.0000 | 1 |
| 10 | 30 | 1 | 1 | 1 | 5.40E−13 | 0.72 | 245.3521 | 2.8534 | 0.9093 | 0.1500 | 0.8465 | 0.3704 | 0.6084 | 18 |
| 11 | 30 | 1 | 2 | 2 | 8.50E−13 | 0.7 | 241.4116 | 3.0980 | 0.8561 | 0.2500 | 0.7766 | 0.4000 | 0.5883 | 19 |
| 12 | 30 | 1 | 3 | 3 | 1.01E−12 | 0.6 | 239.9383 | 4.4370 | 0.8292 | 0.7500 | 0.7454 | 0.6667 | 0.7060 | 9 |
| 13 | 30 | 2 | 1 | 2 | 4.46E−13 | 0.72 | 247.0088 | 2.8534 | 0.9254 | 0.1500 | 0.8702 | 0.3704 | 0.6203 | 17 |
| 14 | 30 | 2 | 2 | 3 | 6.25E−13 | 0.69 | 244.0824 | 3.2230 | 0.8947 | 0.3000 | 0.8261 | 0.4167 | 0.6214 | 16 |
| 15 | 30 | 2 | 3 | 1 | 7.97E−13 | 0.58 | 241.9709 | 4.7314 | 0.8652 | 0.8500 | 0.7877 | 0.7692 | 0.7785 | 6 |
| 16 | 30 | 3 | 1 | 3 | 2.58E−13 | 0.7 | 251.7748 | 3.0980 | 0.9577 | 0.2500 | 0.9220 | 0.4000 | 0.6610 | 13 |
| 17 | 30 | 3 | 2 | 1 | 4.23E−13 | 0.65 | 247.4732 | 3.7417 | 0.9294 | 0.5000 | 0.8762 | 0.5000 | 0.6881 | 11 |
| 18 | 30 | 3 | 3 | 2 | 6.55E−13 | 0.55 | 243.6705 | 5.1927 | 0.8895 | 1.0000 | 0.8190 | 1.0000 | 0.9095 | 2 |

| | | | | | | | | | | | | | | |
|---|---|---|---|---|---|---|---|---|---|---|---|---|---|---|
| 19 | 50 | 1 | 1 | 1 | 1.40E−12 | 0.75 | 237.0774 | 2.4988 | 0.7618 | 0.0000 | 0.6773 | 0.3333 | 0.5053 | 26 |
| 20 | 50 | 1 | 2 | 2 | 1.52E−12 | 0.73 | 236.3460 | 2.7335 | 0.7407 | 0.1000 | 0.6585 | 0.3571 | 0.5078 | 25 |
| 21 | 50 | 1 | 3 | 3 | 5.84E−12 | 0.65 | 224.6688 | 3.7417 | 0.0000 | 0.5000 | 0.3333 | 0.5000 | 0.4167 | 27 |
| 22 | 50 | 2 | 1 | 2 | 1.10E−12 | 0.71 | 239.1815 | 2.9748 | 0.8135 | 0.2000 | 0.7283 | 0.3846 | 0.5565 | 21 |
| 23 | 50 | 2 | 2 | 3 | 1.24E−12 | 0.71 | 238.1056 | 2.9748 | 0.7886 | 0.2000 | 0.7029 | 0.3846 | 0.5437 | 24 |
| 24 | 50 | 2 | 3 | 1 | 2.78E−12 | 0.62 | 231.1164 | 4.1522 | 0.5250 | 0.6500 | 0.5128 | 0.5882 | 0.5505 | 23 |
| 25 | 50 | 3 | 1 | 3 | 1.03E−12 | 0.7 | 239.7517 | 3.0980 | 0.8254 | 0.2500 | 0.7412 | 0.4000 | 0.5706 | 20 |
| 26 | 50 | 3 | 2 | 1 | 1.19E−12 | 0.7 | 238.5075 | 3.0980 | 0.7983 | 0.2500 | 0.7125 | 0.4000 | 0.5563 | 22 |
| 27 | 50 | 3 | 3 | 2 | 2.03E−12 | 0.6 | 233.8501 | 4.4370 | 0.6538 | 0.7500 | 0.5909 | 0.6667 | 0.6288 | 14 |

**Table 10.2** Response table.

| Parameters | Level L1 | L2 | L3 |
|---|---|---|---|
| Load | **0.7711** | 0.6868 | 0.5374 |
| SV | 0.6062 | 0.6653 | **0.7238** |
| % graphite | 0.6178 | 0.6331 | **0.7444** |
| Error | 0.6654 | **0.6878** | 0.6420 |
| Average mean: **0.6122** | | | |

**Table 10.3** ANOVA for various input parameters.

| Parameters | Degrees of freedom | Sum of square | Mean square | Contribution of input parameters (%) |
|---|---|---|---|---|
| Load | 2 | 0.0069 | 0.0034 | 9.9111 |
| Sliding velocity | 2 | 0.0312 | 0.0156 | 44.9881 |
| % graphite | 2 | 0.0305 | 0.0153 | 43.9915 |
| Error | 20 | 0.0008 | 0.0000 | 1.1093 |
| Total | 26 | 0.0694 | — | — |

Gray relational order is found by ranking the gray relational coefficient. The order indicates the best optimal combinations of input parameter to get better output response.

Table 10.2 shows the response table for the input parameters, which indicates that optimal combination of input parameters are load at 10 N, sliding velocity at 3 $m\,s^{-1}$, and reinforcement of graphite with 3%. Table 10.3 shows the ANOVA results; it is obvious that sliding velocity (SV) has the major contribution of 44.99% followed by percentage reinforcement and load with contribution of 43.99% and 9.91% in influencing the output responses wear rate and CoF for the studied composite. The contribution due to error is very minimal compared to other responses, which clearly shows that the achieved result is appropriate.

## 10.4 Conclusion

Composite with fly ash and graphite as reinforcement was successfully fabricated through powder metallurgy technique and the following conclusions have been drawn:

- With increase in load, the wear rate and CoF increase.
- The CoF decreases with increase in sliding velocity and graphite content.
- Sliding velocity has a high contribution in affecting the output responses with 44.99%.

## References


1 Alrashdan, A., Mayyas, A.T., and Hayajneh, M.T. (2011). Drilling of Al–Mg–Cu alloys and Al–Mg–Cu/SiC composites. *Journal of Composite Materials* 45 (20): 2091–2101.

2 Chen, S., Liu, Y., Liu, C., and Sun, G. (2009). Nano-$Al_2O_3$ particle reinforced Cu-based self-lubricating composites. *Acta Materiae Compositae Sinica* 26 (6): 109–115.

3 Hassan, A.M., Alrashdan, A., Hayajneh, M.T., and Mayyas, A.T. (2009). Wear behavior of Al–Mg–Cu-based composites containing SiC particles. *Tribology International* 42 (8): 1230–1238.

4 Pappacena, K.E., Johnson, M.T., Wang, H. et al. (2010). Thermal properties of wood-derived copper–silicon carbide composites fabricated via electrodeposition. *Composites Science and Technology* 70 (3): 478–484.

5 Bhaskar, H.B. and Sharief, A. (2012). The optimization of sliding wear behaviour of aluminium/$Be_3Al_2$ $(SiO_3)_6$ composite using Taguchi approach. *International Journal of Engineering Research and Applications* 2 (1): 64–67.

6 Ghorbani, M., Mazaheri, M., and Afshar, A. (2005). Wear and friction characteristics of electrodeposited graphite–bronze composite coatings. *Surface and Coatings Technology* 190 (1): 32–38.

7 Li, W., Liu, L., and Shen, B. (2011). The fabrication and properties of short carbon fiber reinforced copper matrix composites. *Journal of Composite Materials* 45 (24): 2567–2571.

8 Murthy, H.C.A. and Singh, S.K. (2015). Influence of TiC particulate reinforcement on the corrosion behaviour of Al 6061 metal matrix composites. *Advanced Materials Letters* 6 (7): 633–640.

9 Venkateswaran, K., Kamaraj, M., and Prasad Rao, K. (2007). Dry sliding wear of a powder metallurgy copper-based metal matrix composite reinforced with iron aluminide intermetallic particles. *Journal of Composite Materials* 41 (14): 1713–1728.

10 Xiao, J.-K., Zhang, L., Zhou, K.-C., and Wang, X.-P. (2013). Microscratch behavior of copper–graphite composites. *Tribology International* 57: 38–45.

11 Xue, Z.W., Wang, L.D., Zhao, P.T. et al. (2012). Microstructures and tensile behavior of carbon nanotubes reinforced cu matrix composites with molecular-level dispersion. *Materials and Design* 34: 298–301.

12 Alman, D.E. and Hawk, J.A. (1999). The abrasive wear of sintered titanium matrix–ceramic particle reinforced composites. *Wear* 225–229 (1): 629–639.

13 Badisch, E., Katsich, C., Winkelmann, H. et al. (2010). Wear behaviour of hardfaced Fe–Cr–C alloy and austenitic steel under 2-body and 3-body conditions at elevated temperature. *Tribology International* 43 (7): 1234–1244.

14 Cui, G., Bi, Q., Zhu, S. et al. (2013). Synergistic effect of alumina and graphite on bronze matrix composites: tribological behaviors in sea water. *Wear* 303 (1–2): 216–224.

15 Lee, H.S., Yeo, J.S., Hong, S.H. et al. (2001). The fabrication process and mechanical properties of $SiC_p$/Al–Si metal matrix composites for automobile air-conditioner compressor pistons. *Journal of Materials Processing Technology* 113: 202–208.

16 Puertas, I. and Luis Perez, C.J. (2003). Modelling the manufacturing parameters in electrical discharge machining of siliconized silicon carbide. *Proceedings of the Institution of Mechanical Engineers, Part B: Journal of Engineering Manufacture* 217: 791–803.
17 Yao, P., Sheng, H., Xiong, X., and Huang, B. (2007). Worn surface characteristics of Cu-based powder metallurgy bake materials for aircraft. *Transactions of the Nonferrous Metals Society of China* 17 (1): 99–103.
18 Abhik, R., Umasankar, V., and Xavior, M.A. (2014). Evaluation of properties for Al–SiC reinforced metal matrix composite for brake pads. *Procedia Engineering* 97: 941–950.
19 Ahmad, F., Lo, S.H.J., Aslam, M., and Haziq, A. (2013). Tribology behaviour of alumina particles reinforced aluminium matrix composites and brake disc materials. *Procedia Engineering* 68 (0): 674–680.
20 Kováčik, J., Emmer, Š., Bielek, J., and Kelešі, L. (2008). Effect of composition on friction coefficient of Cu–graphite composites. *Wear* 265 (3–4): 417–421.
21 Muterlle, P.V., Cristofolini, I., Pilla, M. et al. (2011). Surface durability and design criteria for graphite–bronze sintered composites in dry sliding applications. *Materials and Design* 32 (7): 3756–3764.
22 Kovalchenko, A.M., Fushchich, O.I., and Danyluk, S. (2012). The tribological properties and mechanism of wear of Cu-based sintered powder materials containing molybdenum disulfide and molybdenum diselenite under unlubricated sliding against copper. *Wear* 290–291: 106–123.
23 Moustafa, S., El-Badry, S., Sanad, A., and Kieback, B. (2002). Friction and wear of copper–graphite composites made with Cu-coated and uncoated graphite powders. *Wear* 253 (7–8): 699–710.
24 Kalinin, G.M. (2002). Specification of properties and design allowables for copper alloys used in HHF components of ITER. *Journal of Nuclear Materials* 311: 668–672.
25 Kato, H., Takama, M., Lwai, Y. et al. (2003). Wear and mechanical properties of sintered copper–tin composites containing graphite or molybdenum disulfide. *Wear* 255: 573–578.
26 Hussain, Z. and Kit, L.C. (2008). Properties and spot welding behaviour of copper–alumina composites through ball milling and mechanical alloying. *Materials and Design* 29 (7): 1311–1315.
27 Hong, E., Kaplin, B., You, T. et al. (2011). Tribological properties of copper alloy-based composites reinforced with tungsten carbide particles. *Wear* 270 (9–10): 591–597.
28 Ram Prabhu, T., Varma, V.K., and Vedantam, S. (2014). Tribological and mechanical behavior of multilayer Cu/$SiC_p$ Gr hybrid composites for brake friction material applications. *Wear* 317 (1–2): 201–212.
29 Rajan, T.P.D., Pillai, R.M., Pai, B.C. et al. (2007). Fabrication and characterisation of Al–7Si–0.35Mg/fly ash metal matrix composites processed by different stir casting routes. *Composites Science and Technology* 67 (15–16): 3369–3377.
30 Daoud, A. (2007). Fabrication, microstructure and compressive behavior of ZC63 Mg–microballoon foam composites. *Composites Science and Technology* 67: 1842–1853.

31 Lu, J., Xu, F., Wang, D. et al. (2009). The application of silicalite-1/fly ash cenosphere (S/FAC) zeolite composite for the adsorption of methyl *tert*-butyl ether (MTBE). *Journal of Hazardous Materials* 165: 120–125.

32 Ramachandra, M. and Radhakrishna, K. (2013). Microstructure, mechanical properties, wear and corrosion behaviour of Al–Si/$FLYASH_P$ composite. *Materials Science and Technology* 21 (11): 1337–1343.

33 Rohatgi, P.K., Guo, R.Q., Huang, P., and Ray, S. (1997). Friction and abrasion resistance of cast aluminum alloy–fly ash composites. 28: 245–250.

34 Bhattacharjee, U. and Kandpal, T.C. (2002). Potential of fly ash utilization in India. *Energy* 27: 151–166.

35 Mahendra, K.V. and Radhakrishna, K. (2009). Characterization of stir cast Al–Cu–(fly ash + SiC) hybrid metal matrix composites. *Journal of Composite Materials* 44 (8): 989–1005.

36 Kumar, P.R.S., Kumaran, S., Rao, T.S., and Natarajan, S. (2010). High temperature sliding wear behavior of press-extruded AA6061/fly ash composite. *Materials Science and Engineering A* 527 (6): 1501–1509.

37 David Raja Selvam, J., Robinson Smart, D.S., and Dinaharan, I. (2013). Microstructure and some mechanical properties of fly ash particulate reinforced AA6061 aluminum alloy composites prepared by compocasting. *Materials and Design* 49: 28–34.

38 Sreekanth, M.S., Bambole, V.A., Mhaske, S.T., and Mahanwar, P.A. (2009). Effect of particle size and concentration of fly ash on properties of polyester thermoplastic elastomer composites. *Journal of Minerals and Materials Characterization and Engineering* 8 (3): 237–248.

39 Sai, N.V., Komaraiah, M., and Raju, A.V.S.R. (2008). Preparation and properties of sintered copper–tin composites containing copper coated or uncoated fly ash. *Materials and Manufacturing Processes* 23 (7): 651–657.

40 Vogiatzis, C.A., Tsouknidas, A., Kountouras, D.T., and Skolianos, S. (2015). Aluminum-ceramic cenospheres syntactic foams produced by powder metallurgy route. *Materials and Design* 85 (1): 444–454.

41 Fan, L.J. and Juang, S.H. (2016). Reaction effect of fly ash with Al–3Mg melt on the microstructure and hardness of aluminum matrix composites. *Materials and Design* 89: 941–949.

42 Rao, J.B., Rao, D.V., Murthy, I.N., and Bhargava, N. (2012). Mechanical properties and corrosion behaviour of fly ash particles reinforced AA 2024 composites. *Journal of Composite Materials* 46 (12): 1393–1404.

43 Sajjadi, S.A., Ezatpour, H.R., and Torabi Parizi, M. (2012). Comparison of microstructure and mechanical properties of A356 aluminum alloy/$Al_2O_3$ composites fabricated by stir and compo-casting processes. *Materials and Design* 34: 106–111.

44 Sapate, S.G., Uttarwar, A., Rathod, R.C., and Paretkar, R.K. (2009). Analyzing dry sliding wear behaviour of copper matrix composites reinforced with pre-coated $SiC_p$ particles. *Materials and Design* 30: 376–386.

45 Narayanasamy, R., Anandakrishnan, V., and Pandey, K.S. (2008). Some aspects on plastic deformation of copper and copper–titanium carbide powder metallurgy composite preforms during cold upsetting. *International Journal of Material Forming* 1 (4): 189–209.

46 Ayyappadas, C., Muthuchamy, A., Annamalai, A.R., and Agrawal, D.K. (2017). An investigation on the effect of sintering mode on various properties of copper–graphene metal matrix composite. *Advanced Powder Technology* 28 (7): 1760–1768.
47 Rajkumar, K. and Aravindan, S. (2011). Tribological studies on microwave sintered copper–carbon nanotube composites. *Wear* 270 (9–10): 613–621.
48 Jia, J., Chen, J., Zhou, H. et al. (2004). Friction and wear properties of bronze–graphite composite under water lubrication. *Tribology International* 37 (5): 423–429.

# 11

# Morphological Examination of Worn out Surfaces of Basalt Fiber-PEI Composites with Varying Loading Conditions

*Kalimuthu Mayandi*[1], *Subramanian Karthikeyan*[2], *Nagarajan Rajini*[1], *and Azeez B. Alavudeen*[1]

[1] *Kalasalingam University, Department of Mechanical Engineering, Anand Nagar, Krishnankoil 626126, Tamil Nadu, India*
[2] *Kalasalingam University, Department of Automobile Engineering, Anand Nagar, Krishnankoil 626126, Tamil Nadu, India*

## 11.1 Introduction

The use of polymer composite materials (PCMs) is increasing in various application areas such as aerospace, marine, construction field, and sports due to its attractive properties. Properties such as weight-to-strength ratio, Young's modulus, elongation, and corrosion resistance are high in PCM compared to traditional materials. Moreover, the fabrication methods and materials handling of PCMs are easy compared to traditional materials. In addition, PCM has light weight and is easily transferable [1–4]. This work studies the basalt fiber wear characteristics for low loading application areas. The basalt fiber withstands high temperatures up to 1700 °C. It is one of the natural resource materials obtained from the lava of volcanic rocks. Among the various natural materials, basalt has high tensile strength and modulus when compared to other natural resource materials such as banana, sisal, pineapple, and bamboo. Glass fibers lead to irritation to the human skin during direct handling. This does not happen with the basalt fiber and it is easy to handle during fabrication [5, 6]. Moreover, large-scale industries consider basalt fibers for reinforcement in polymer matrix, which can find application in heavy-duty friction units requiring high strength. Further, basalt can replace the carbon fiber in areas of high temperature application due to economic considerations and availability [7].

In an earlier study, the mechanical performance of basalt fiber reinforced PCMs was found to be superior to that of glass fiber reinforced composites materials [8]. Tapi et al. [9] investigated that basalt with thermoplastic has improved mechanical and wear properties. Polyetherimide (PEI) is one of the thermoplastic materials. It has good tensile strength and modulus properties. It can be used in various engineering application areas such as electrical and electronics parts, microwave applications, and automobile industries. PEI is one of the newest engineering thermoplastics materials [10]. Tiwari et al. [11]

*Synthesis and Tribological Applications of Hybrid Materials*, First Edition.
Edited by Mohammad Jawaid, Rajini Nagarajan, Jacob Sukumaran, and Patrick De Baets.

developed the composites materials with polyetherimide as a matrix for a study of the tribological and mechanical properties. They concluded that the composite materials developed with PEI had better properties in all aspects. Polytetrafluroethylene (PTFE) is one of the filler materials for polymer matrix composite materials. It has high mechanical performance and tribological properties. The PTFE filled materials have better improvement in wear resistance compare to unfilled composite materials [12]. The recently developed jute fiber hybrid composites are filled with PTFE as the filler material. It was observed that the addition of PTFE in polymer composites showed better performance in high loading wear application areas [12, 13]. The influence of reinforcement with some synthetic fibers such as glass, carbon, and kevlar on PTFE filled composites has been studied for various tribological aspects with respect to different testing parameters [14–16]. Similarly, there are many reports on the tribological behavior of PTFE filled composites with carbon fiber as reinforcement in various thermoplastic systems that were found to exhibit better frictional and wear behavior.

This chapter discusses the friction and wear properties of basalt/PEI composites with and without the addition of PTFE and comparison is made with the results for glass/PEI composites. The same tribology testing conditions for glass/PEI composite have been used in the case of basalt/PEI seen in literature. A hot compression technique has been used for the fabrication of basalt/PEI composites.

## 11.2 Materials Used

Plain woven basalt fabric was used as the primary reinforced material. It was procured from the Nickunj Eximp Enterprises Pvt Ltd, Chennai, India. The fabric was obtained in the form of a roll. A solution of PEI and dichloromethane was used as the matrix material. These materials were obtained from Sigma Aldrich (P) Ltd, Bangalore, India. In addition, PTFE was selected as filler material to reduce the coefficient of friction (CoF).

## 11.3 Fabrication of the Composite Materials

Untreated basalt fabric, PEI, and PTFE were used to fabricate the composite materials. First, the basalt fabric was cut into the required size of 100 mm × 100 mm and the PEI solution was prepared by adding 25 wt% of dichloromethane. The prepared solution was kept in a container vessel. Then, the basalt fabric was immersed in the container PEI solution for 1 h followed by drying at atmospheric room temperature. Twenty one prepergs were prepared by using the PEI solution. These prepergs were used to fabricate a composite plate of 3 mm thickness. All the prepergs were placed in the cavity of a bottom mold. The top cover of the mold was kept closed and placed in the hot compression machine and the compression pressure was set at 15 MPa and temperature at 350 °C. A second specimen was also fabricated in the same manner with the addition of 10 wt% of PTFE as filler material in PEI solution.

## 11.4 Testing of Composite Materials

### 11.4.1 Density Test

The density of the prepared composite was measured using Mettler Toledo densitometer. It was used with water as the immersing liquid. Five samples were tested for each condition in the ASTM D792 standard. The average density of the basalt/PEI composite was 1.0052 $\mathrm{g\,cm^{-3}}$. The density of the same composite with the addition of PTFE filler increased to 1.0086 $\mathrm{g\,cm^{-3}}$.

### 11.4.2 Hardness Test

Hardness measurements were carried out on 3 mm thick specimens as per ASTM D2240 standards on a shore D scale. Indentations were made at several locations for each specimen and the average hardness value was calculated. At least five specimens were tested and the average value of the basalt/PEI composite was measured as 98.06. The addition of filler materials to the composite caused a marginal improvement in the hardness up to 99.82.

### 11.4.3 Wear Test

In the present work, the friction and dry sliding wear behavior of basalt/PEI and basalt/PEI/PTFE composite samples have been studied in terms of the coefficient of friction and specific wear rate. A pin-on-disk setup (as per ASTM G-99 standard, Make: Magnum Engineers, Bangalore) was used for the sliding wear and two-body wear tests. The surface of an 8 mm diameter composite specimen, glued to a pin of 10 mm width and 50 mm length, comes in contact with a hardened alloy steel disk with hardness value of 62 HRC and surface roughness (Ra) of 0.54 μm. The test was conducted on a track of 50 mm diameter for a specified test duration, load, and velocity. The specimen was initially weighed using a digital electronic balance (0.1 mg accuracy). The test was carried out by applying normal load (70, 80, 90, 100 N) and run for a constant sliding distance of 1413 m at 150 rpm. At the end of the test, the sample was again weighed in the same balance. The difference between the initial and final mass was used as a measure of wear loss. The sample was placed so that the specimen was perpendicular to the steel disk and parallel with respect to the abrading direction. All the tests were conducted at ambient temperature. The specific wear rate was calculated using the following relation:

$$\mathrm{Ko} = \Delta V / L \times D\,\mathrm{m^3\,N^{-1}\,m^{-1}} \tag{11.1}$$

where Ko is the specific wear rate, $\Delta V$ is the volume loss, $L$ is the load, and $D$ is the sliding distance.

## 11.5 Results and Discussion

### 11.5.1 Wear Performance of Basalt Fiber Reinforced Thermoplastic Composite

The coefficient of friction of basalt/PEI and basalt/PEI/PTFE composite for different loading and sliding conditions was plotted as a function of test duration. It was observed that the CoF ($\mu$) increased slightly with increasing load for the basalt/PEI composite. Figures 11.1 and 11.2 clearly show the appearance of these variations in wear within the range and also the positive behavior of the material toward wear resistance. It was found that almost both the composite materials attained a stable state of friction in 15 min of sliding.

Inclusion of PTFE in the basalt reinforced composite, however, affects the CoF ($\mu$) in an unfavorable way. A coefficient of friction ($\mu$) value as low as 0.14 was reached in the basalt/PEI composite, but, in the case of the PTFE filled composite, it was 0.284. However, for 80 and 90 N loading conditions, there was no significant difference in friction characteristics between the two composite materials. It was clearly noticed that with the increase in sliding duration and loading conditions, $\mu$ changed very slowly. The frictional value of basalt fabric is compared with that of glass fiber from the literature. The $\mu$ value of the both the composites is found to be very low when compared to glass/PEI and PTFE filled glass/PEI composite. The addition of PTFE in PEI changes the contact condition with the counter surface and increases the overall friction value slightly from 0.23 to 0.3. The basalt/PEI composite shows better performance in terms of CoF.

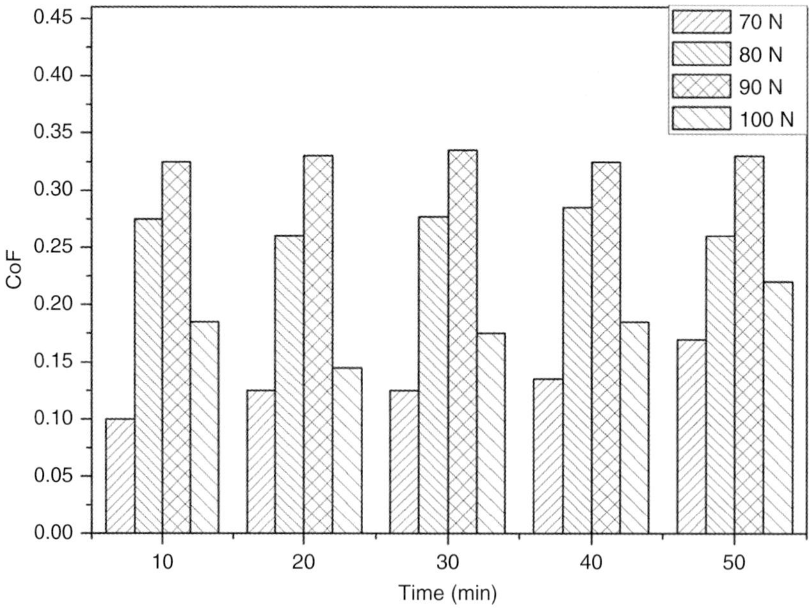


**Figure 11.1** Coefficient of friction of basalt/PEI.

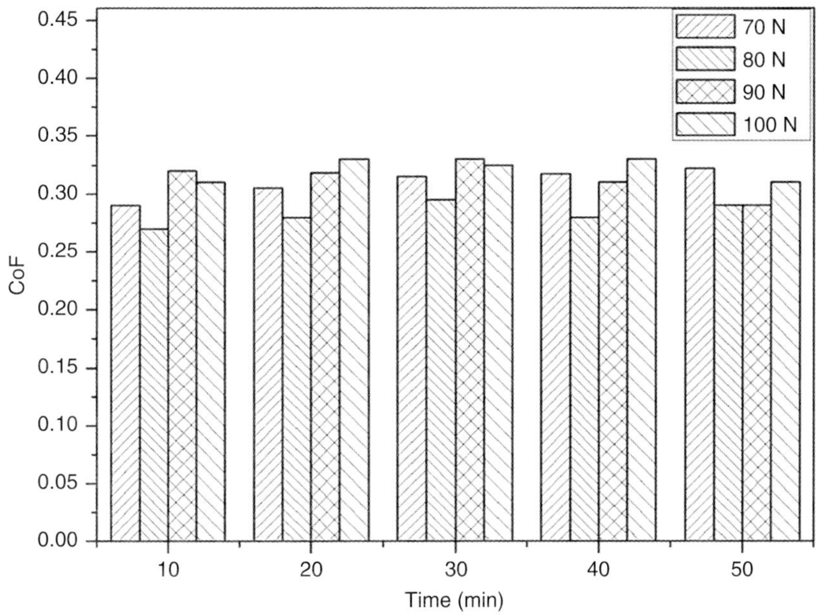


**Figure 11.2** Coefficient of friction of basalt/PEI/PTFE.

Selection of the same tribological test configuration and load from the existing literature allows comparison of the present result with the reported data. It is evident from Figures 11.3 and 11.4 that there is a significant improvement in the friction characteristics of the basalt-PEI composite when compared with the glass fiber PEI composite. Similar results are observed when comparing the

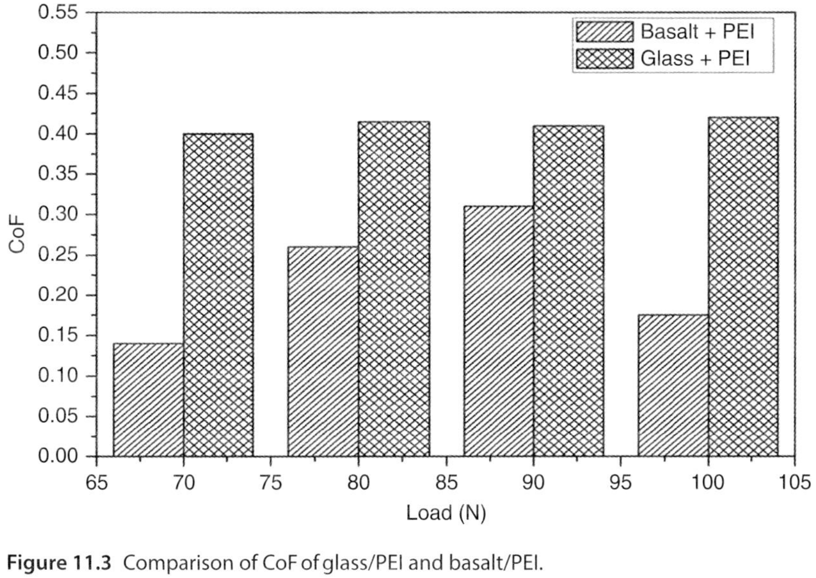


**Figure 11.3** Comparison of CoF of glass/PEI and basalt/PEI.

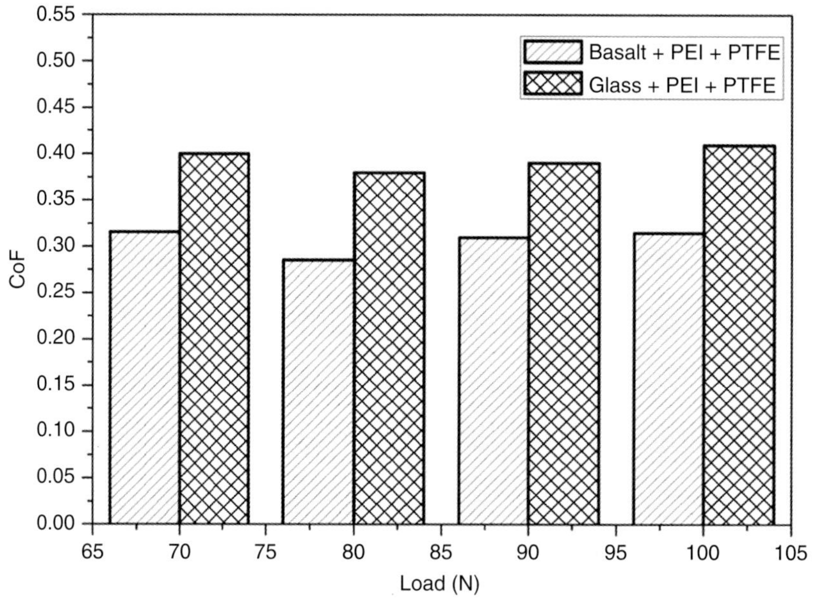


**Figure 11.4** Comparison of CoF of basalt/PEI/PTFE and glass/PEI/PTFE.

influence of the PTFE addition to the same composite, where the basalt fiber composite with PTFE additives has lower friction values compared to the counterparts in the literature. Both the composites exhibit wear rates in the order of $10^{-15}$ $m^3$ $N^{-1}$ $m^{-1}$. A similar tendency in the friction characteristics is observed for both the basalt-PEI and basalt-PEI-PTFE composites. At higher loads, the PTFE composites showed increased wear rate.

Generally, increase in load can increase the magnitude of friction force and therefore the contact temperature could also increase at the contact surface. After a certain temperature, depletion of materials takes place and it adheres to the counter surface. Moreover, a continuous increase in temperature can remove the matrix component as much as possible. It implies that the glass transition temperature could have appeared in the contact surface, so that molecule disorder can be initiated in the PEI molecular structure and propagate further with increasing temperature. At the same time, an almost constant specific wear rate was observed for both the composites with the effect of varying load. However, a significant reduction in specific wear rate was noticed for basalt/PTFE composites. The results show the removal of PTFE to be higher in the case of glass/PEI composition. It could happen due to the non-occurrence of re-adhesion of PTFE sample. But in the case of basalt/PTFE composite, the adhesion of PTFE on the counterface could be strong enough to ensure smooth running action. Further, the surface of PTFE transfer layer on the counter plate can be the reason for achieving this lower specific wear rate.

Figure 11.3 shows the comparative results of CoF for glass/PEI and basalt/PEI composites with the application of varying loads. They reveal almost consistent

CoF in the case of glass/PEI composites. On the other hand, an irregular variation in CoF was noticed in basalt/PEI composites. However, the value of CoF is always lower than the values of glass/PEI composites. It was realized from the results that the compatibility of PEI with glass and basalt fiber may vary due to different morphologies of the fiber surface.

The results show that the degree of adhesion between the basalt and PTFE is higher than that of glass/PEI composition. This is confirmed from the morphological analysis of basalt/PEI worn out samples. This is discussed in the forthcoming sections. The removal of PEI matrix after the running-in period of testing condition can expose the fiber to contact with the counter surface.

A similar kind of variation in CoF was observed in both the composites with the addition of PTFE filler as seen in Figure 11.4. CoF was seen decreasing from 70 N load to 80 N load, clearly indicating the transfer of material occurring from the sample to the counterface at higher loading conditions. Later, the adhesion and re-adhesion could have happened with increasing load.

### 11.5.2 Morphological Analysis of Worn out Samples

Microscopic analysis was performed for two different extreme loading conditions at 70 and 100 N with and without basalt/PTFE addition for examining the wear behavior of basalt/PTFE composites as shown in Figure 11.5a,b. Figure 11.5a clearly shows the morphology of PTFE at 70 N load. The wear track along the direction of friction force clearly indicates the absence of any material removal. However, a smooth surface was found with the partially melted and scratched surface and re-adhered particle. In the case of PTFE coated material (Figure 11.5b), large removal of PTFE was seen clearly in a few thin patches, which led to the discontinuous surface causing an increase in the friction force. As expected, the detached particles from the samples were found to be adhered firmly to the counter surface. Hence, a steady state friction was achieved in the case of 100 N loading condition after a certain period of time.

Figure 11.6a,b shows the topography of basalt/PEI worn out composites with and without PTFE addition at lower loading (70 N) conditions. Figure 11.6a shows only a straight longitudinal scratch on the surfaces of uncoated composites. Further, the pressure of wear debris with small dimensions is also seen from the surface. This is responsible for the scratch along the normal to the frictional direction. In Figure 11.6b the removal of PTFE is noticed on the surface, indicating the insufficient bonding of PTFE particle with the highly viscous thermoplastic PEI. However, the tribo lubricant transfer to the counterpart can reduce the magnitude of frictional force until the existence of transfer layer on the counter surface.

The higher magnified image was compared at 500× magnification for basalt/PEI composites under 70 N loading conditions to get a close look at the worn out surfaces. Figure 11.7a shows that the depleted PEI matrix occupies the space between the fibers and forms the smooth and even surface of

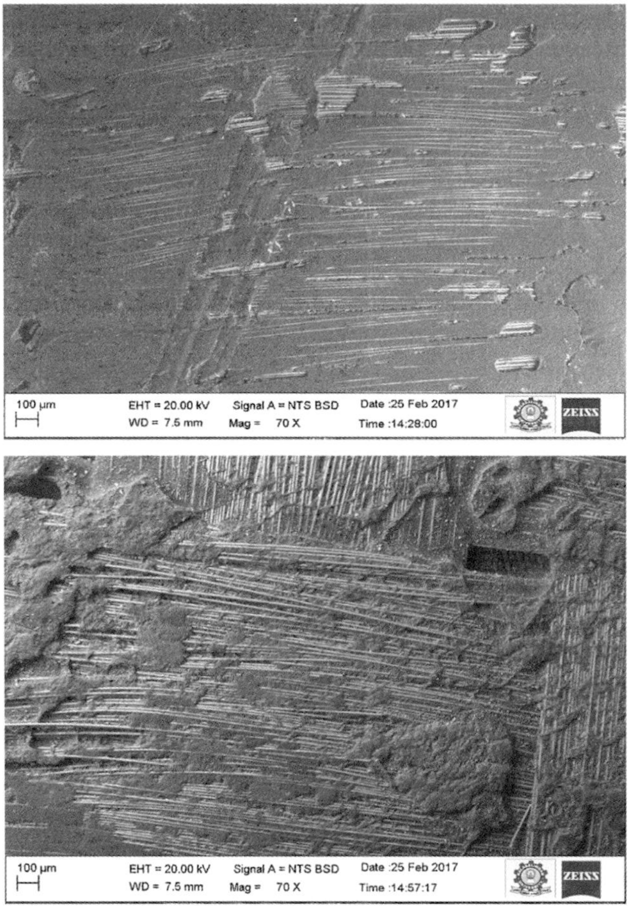


**Figure 11.5** (a) SEM micrograph of basalt/PEI composite and (b) with PTFE.

the samples. This can minimize the changes of asperity content between the materials. At the same time, the larger gap between the fibers shown in Figure 11.7b indicates the adherence of PTFE coated matrix on the counter plate. Further, the exposure of fiber seems to be dominant compared to the previous case.

At a higher loading condition, large quantity matrix removal was noticed. This is shown in Figure 11.8a. Further, fiber breakage, interfacial debonding, and separation of loosely bonded wear debris with small particle size were observed at the interface between the fibers. It implies that the increase in friction force due to the application of higher load ensures adherence and re-adherence of the matrix on the counter plate. In Figure 11.8b, micro-ploughing with deeper grooves is seen in the case of PTFE coated composites. It confirms the irregular adhesion of PTFE coating on the counter surface and thus it creates a higher magnitude of asperity region at the contact surface.

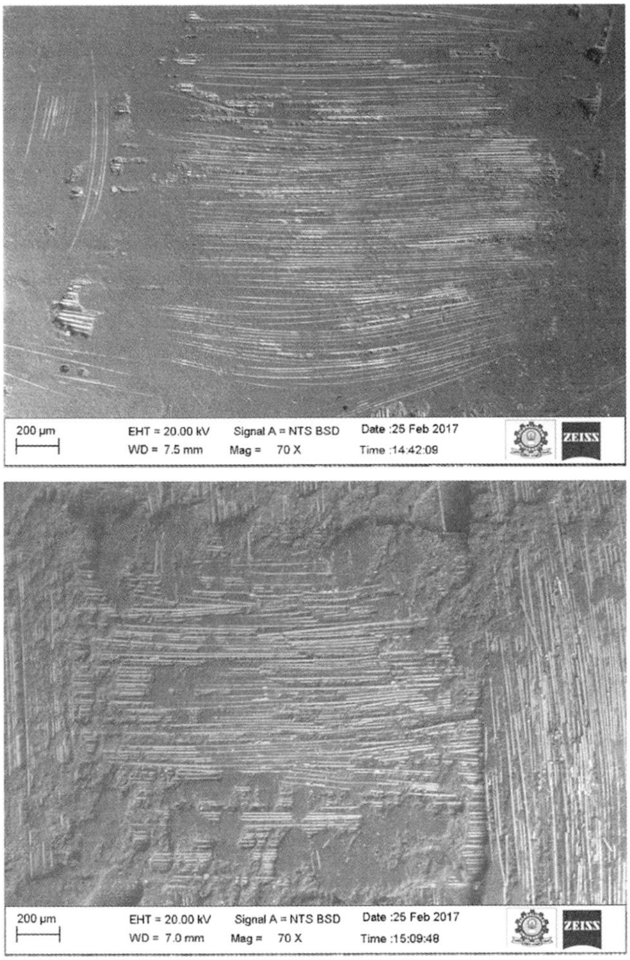


(a)

(b)

**Figure 11.6** (a) SEM micrograph of basalt/PEI composite at 70× and (b) with PTFE at 70×.

## 11.6 Conclusions

An attempt has been made in the research work to introduce PEI basalt fiber composites. From the investigation on mechanical and tribological testing the following conclusions are drawn.

- The hardness of the basalt-PEI composites remained the same with the addition of PTFE.
- CoF values are lower for basalt PEI composite when compared with the basalt PEI-PTFE composite.
- However, at 80 and 90 N loading condition no significant change was observed in the friction characteristics.

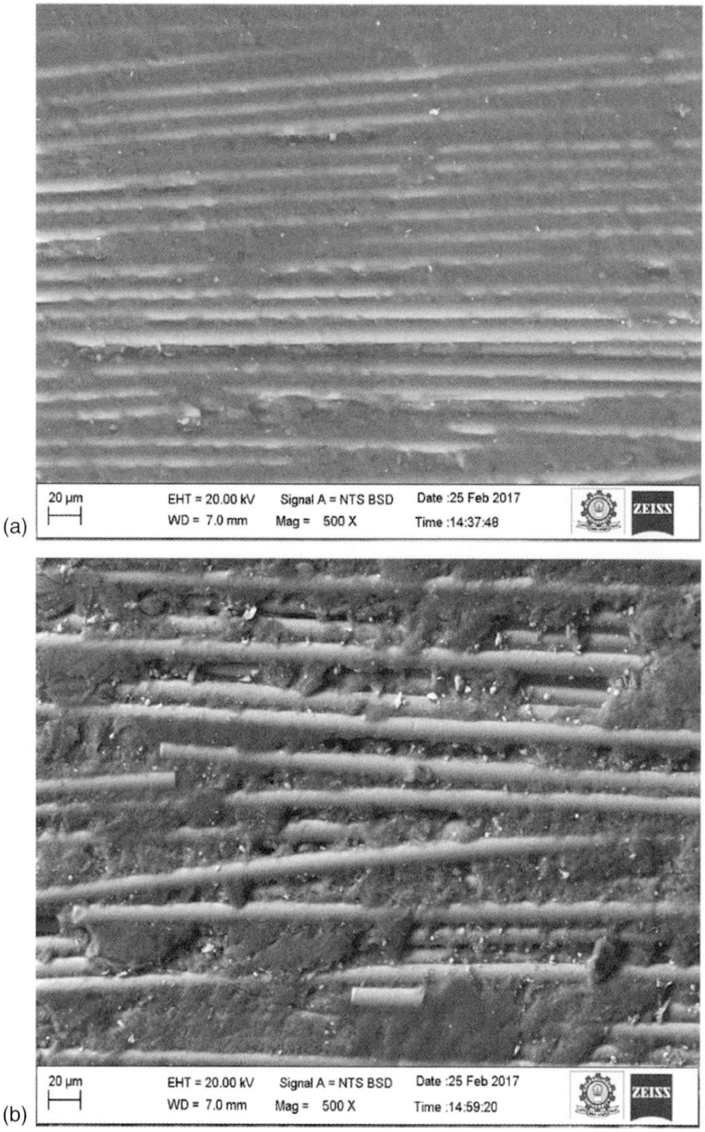


**Figure 11.7** (a) SEM micrograph of basalt/PEI composite at 500× and (b) with PTFE at 500×.

- The wear rate showed similar characteristics as the friction behavior where the basalt/PEI composite has better wear resistance when compared with the basalt PEI-PTFE composite.
- Morphological analysis confirmed that after wear test the basalt/PEI composite sample has better smooth surfaces while the worn out surfaces of basalt/PEI/PTFE composites materials have more rough surfaces.
- On comparison with the existing literature, the newly developed polymer stands out as a suitable candidate material for tribological applications.

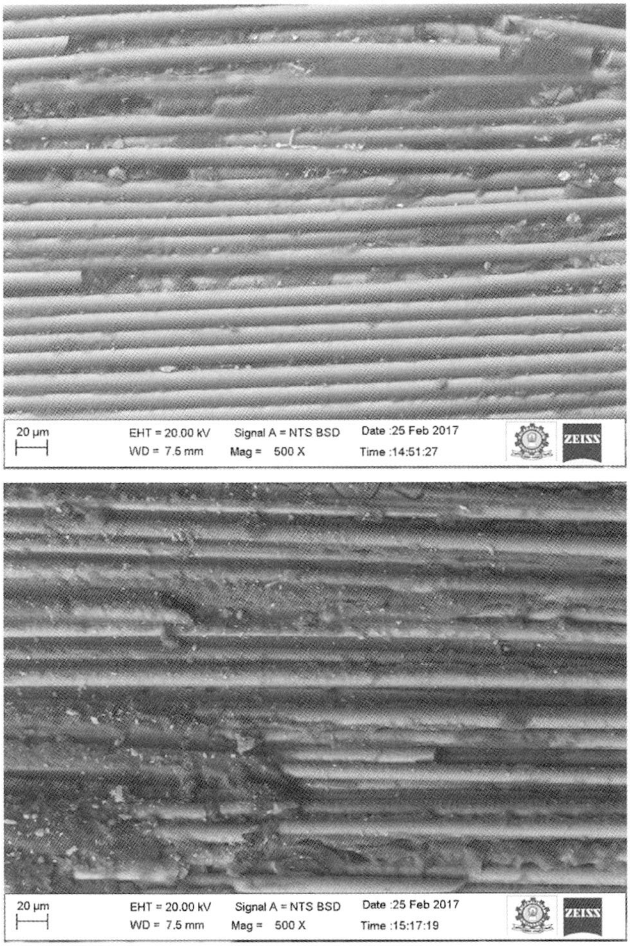


**Figure 11.8** (a) SEM micrograph of basalt/PEI composite and (b) with PTFE.

## References

1 Wambua, P. and Ivens, J. (2003). Natural fibres: can they replace glass in fibre reinforced plastics. *Composites Science and Technology* 63 (16): 1259–1267.
2 Sreenivasan, V.S., Ravindran, D., Manikandan, V., and Narayanasamy, R. (2011). Mechanical properties of randomly oriented short *Sansevieria cylindrica* fibre/polyester composites. *Materials and Design* 32: 2444–2455.
3 Obi Reddy, K., Ashok, B., Narender Reddy, K.R. et al. (2014). Extraction and characterization of novel lignocellulosic fibers from *Thespesia lampas* plant. *International Journal of Polymer Analysis and Characterization* 19: 48–61.
4 Mayandi, K., Rajini, N., Pitchipoo, P. et al. (2015). Properties of untreated and chemically treated Cissus quadrangularis natural fibers and their composites with polyester as the matrix. *Materials Today Communications* doi: 10.1002/pc.24011.

5 Chairman, C.A. and Kumaresh Babu, S.P. (2013). Mechanical and abrasive wear behavior of glass and basalt fabric-reinforced epoxy composites. *Journal of Applied Polymer Science* 130 (1): 120–130.
6 Kogan, F.M. and Nikitina, O.V. (1994). Solubility of chrysotile asbestos and basalt fibers in relation to their fibrogenic and carcinogenic action. *Environmental Health Perspectives* 102: S205–S206.
7 Zhang, X., Pei, X., and Wang, Q. (2010). Friction and wear behavior of basalt-fabric-reinforced/solid-lubricant-filled phenolic composites. *Journal of Applied Polymer Science* 117: 3428–3433.
8 Manikandan, V., Winowlin Jappes, J.T., Suresh Kumar, S.M., and Amuthakkannan, P. (2012). Investigation of the effect of surface modifications on the mechanical properties of basalt fibre reinforced polymer composites. *Composites Part B: Engineering* 43 (2): 812–818.
9 Tábi, T., Tamás, P., and Kovács, J.G. (2013). Chopped basalt fibres: a new perspective in reinforcing poly(lactic acid) to produce injection moulded engineering composites from renewable and natural resources. *Express Polymer Letters* 7 (2): 107–119.
10 Bijwe, J., Tewari, U.S., and Vasudevan, P. (1989). Friction and wear studies of a short glass-fibre-reinforced polyetherimide composite. *Wear* 132: 247–264.
11 Tiwari, S. and Bijwe, J. (2013). Various ways to strengthen the fiber-matrix interface for enhanced composite performance. *Surface and Interface Analysis* 45 (13): 1838–1848.
12 Khedkar, J., Negulescu, I., and Meletis, E.I. (2002). Sliding wear behavior of PTFE composites. *Wear* 252: 361–369.
13 Subramanian, K., Nagarajan, R., Sukumaran, J. et al. (2015). Dry sliding wear properties of jute/polymer composites in high loading applications. *Mechanical Engineering Letters* 12: 7–18.
14 Briscoe, B.J., Steward, M.D., and Groszek, A.J. (1977). The effect of carbon aspect ratio on the friction and wear of PTFE. *Wear* 42: 99–107.
15 Lhymn, C. (1986). Microscopic study of the frictional wear of PTFE. *Wear* 107: 95–105.
16 Zhang, Z., Xue, Q.J., Liu, W. et al. (1998). Friction and wear characteristics of lead and its compounds filled polytetrafluoroethylene composites under oil lubricated conditions. *Tribology International* 31: 361–368.

# Index



*Synthesis and Tribological Applications of Hybrid Materials*, First Edition.
Edited by Mohammad Jawaid, Rajini Nagarajan, Jacob Sukumaran, and Patrick De Baets.

## y